Lecture Notes in Control and Information Sciences

Edited by M. Thoma and A. Wyner

Lecture Notes in Control and Information Sciences

Edited by M. Thoma and A. Wyner

167

M. Rao

Integrated System for Intelligent Control

Springer-Verlag
Berlin Heidelberg GmbH

Author

Prof. Ming Rao
Intelligence Engineering Laboratory
Dept. of Chemical Engineering
University of Alberta
Edmonton, Alberta
Canada T 6G 2G6

ISBN 978-3-540-54913-0   ISBN 978-3-540-46501-0 (eBook)
DOI 10.1007/978-3-540-46501-0

Typesetting: Camera ready by author

61/3020-543210 Printed on acid-free paper.

PREFACE

Intelligent control is a new interdisciplinary field which extensively applies the knowledge of computer science, artificial intelligence, electrical engineering as well as system science to industrial automation processes.

A new integration architecture for implementing real-time distributed intelligent control systems is also developed. The construction of intelligent systems is one of the most important techniques among artificial intelligence research tasks. My goal is to develop an integrated intelligent system to accomplish the real-time control of industrial processes. An integrated intelligent system is a large knowledge integration environment that consists of several symbolic reasoning systems (expert systems) and numerical computation packages. These software programs are controlled by a meta-system, which manages the selection, operation and communication of these programs. This new architecture can serve as a universal configuration to develop high-performance distributed intelligent systems for many complicated applications in real-world domains. The configuration of the integrated intelligent system has attracted significant attention from both industry and academia, and is expected to lead to a new era for the application of AI techniques to real-world chemical intelligent process control problems.

My experience from developing intelligent process control systems also indicates that new knowledge may be generated from the process of developing knowledge-based systems, which may complement the knowledge of both artificial intelligence techniques and the related application domains.

Chapter 1 introduces the background of intelligent control system, its current state, and future development. A knowledge-based systems for process control design, namely IDSCA (Intelligent Direction Selector for the Controller's Action in multiloop control systems), is presented in Chapter 2. The integrated intelligent system is described in Chapter 3. Its implementation in OPS5 environment and C environment is presented the following chapter, while the implementation with TURBO-PROLOG is presented in Chapter 8 with application to an intelligent gear manufacturing system. Several applications based on the integrated intelligent system are developed, such as intelligent optimal control system (Chapter 5), pulp and paper intelligent process control (Chapter 6), intelligent maintenance support system for air-traffic control (Chapter 7), and integrated intelligent software environment for gear manufacturing system (Chapter 8). The new knowledge, which is generated form building intelligent control systems and may complement the knowledge of both AI technique and application domains, is covered in Chapter 9. The conclusions are summarized in Chapter 10.

I would like to express my appreciation towards Dr. Charles Theisen and Dr. Murray Wonham, who give me the important advice and suggestions, strong encouragement and support to prepare this monograph. I would like to appreciate Dr. Thoma, the editor of Springer-Verlarg Lecture Notes in Control and Information Sciences Series, for his very valuable suggestions to revise this manuscript.

The graduate students (Jean Corbin, Randy Dong, Heon-Chang Kim, Murray Stevenson, Yiqun Ying and Hong Zhou), postdoctoal fellows (Jiangzhong Cha, Pin-Chan Du, Xuemin Shen, Qun Wang and Qijin Xia) and research associates (Haiming Qiu) who work in my laboratory have made the important contributions to the technical contents as well as the preparation of this manuscript. My colleague and friends (Tsung-Shann Jiang, Jeffrey J.P. Tsai, James Luxhoj, Shaw Wang, Guohong Wu, Rafael Cruz, Ji Zhou, and Grantham Pang) provide the necessary assistance to me. Thanks are also due to my wife Xiaomei Zheng and Mr. Henry Sit for their help in preparing this monograph.

I gratefully acknowledge for the financial support from Natural Sciences and Engineering Research Council of Canada, the National Science Foundation (USA), the University of Alberta, and Rutgers University.

TABLE OF CONTENTS

NOMENCLATURE

AC	Automatic Control
AI	Artificial Intelligence
AFTS	Adaptive Feedback Testing System
CAD	Computer-Aided Design
CACSD	Computer-Aided Control Systems Design
CAM	Computer-Aided Manufacturing
CAT	Computer-Aided Testing
CIMS	Computer Integrated Manufacturing System
D	disturbance
DB	database
E	error signal, defined as X - Z
ESID	Expert of System Identification Design
G	transfer function
GIMS	Gear Integrated Manufacturing System
ICAD	Intelligent Computer-Aided Design
ICAM	Intelligent Computer-Aided Manufacturing
ICAT	Intelligent Computer-Aided Testing
IDSCA	Intelligent Direction Selector for Controllers' Action
IDSOC	Intelligent Decisionmaker for Solving Optimal Control
IIS	Integrated Intelligent System
IMSS	Intelligent Maintenance Support System
LHS	LIFT-HAND-SIDE
KB	knowledge base
KBS	Knowledge-Based System
K_p	process gain
NC	numerical control
OOP	object-oriented programming
P	output signal of the controller
PID	Proportional, Integral and Derivative
PLC	Programming Logic Controller
Q	manipulative variable
R	action direction function
RHS	RIGHT-HAND-SIDE
T_d	differential time
T_i	integral time
U	control variable
X	set point for the controller
Y	controlled process output variable
Z	measured value of Y

Subscript

b	block
c	controller
(e(k))	a sequence of independent normally distributed random variables.
i	the ith loop ($1 < i < n$)
l	loop
m	measuring device
n	inner most control loop
p	process
q	disturbance magnitude
v	control valve
1	primary control loop (outer most loop)
2	secondary control loop (inner loop)

1 Introduction

1.1 Intelligent control for industrial processes

The growing complexity of industrial processes and the need for higher efficiency, greater flexibility, better product quality, and lower cost have changed the face of industrial practice. Meanwhile, the application of computers has allowed the implementation of more advanced techniques. For example, chemical process control is the science and technology of automation in chemical industries. As shown in Figure 1.1, chemical process control is interdisciplinary in nature, and allows the application of knowledge from system science and computer science to be extensively applied.

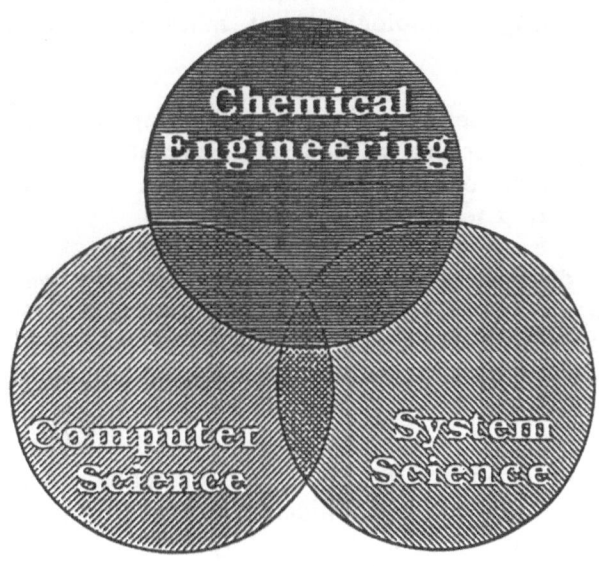

Figure 1.1 Field interaction

It has been widely recognized that quality control and process automation are the key elements to make modern industries stay competitive internationally. Recently, the chemical companies have begun to recognize the importance of process control in order to have a successfully functioning manufacturing facility [Astrom, 1985]. With the advent of microcomputer technology, process control developed rapidly during the past 10 years. The chemical process industries have historically recognized the importance of process control in order to have a successfully functioning manufacturing facility. It can safely be said that chemical processing in the future will require that operators have considerable process control knowledge and experience.

For a long time, computers have been used for control and monitoring in process industries. With the advent of microcomputer technology, process control developed rapidly. The application of more powerful computers has allowed us to implement more advanced control concepts. The systematic knowledge of process operation has created an environment facilitating the introduction of expert systems. Existing hardware and programming technologies have matured [Ishii and Hayami, 1988].

On the other hand, the increasing demand for more effective data processing methods and control strategies to be used in a wider range of industrial applications can not be met by current computing techniques.

The development of industry automation may be divided into four stages in term of automation [Lu, 1989]. The first stage, namely labor-intensive stage, mainly relies on skills of human operators on simple non-automatic machines, which can be designed by individual designer. At the second stage (equipment-intensive), automatic equipment plays a dominant role in the competition of productivity. This kind of machines are associated with the complicated mechanical, electronic and computerized devices which can only be designed by a group of human design experts from different domains. As a result of more powerful and affordable computing facilities on factory floors, our industry is now moving into the third stage (data-intensive), which stimulates the development of Flexible Manufacturing System (FMS). At this stage, automation is realized at the level of data processing, and CAD is the key technology for design tasks. The next challenge is decision-making automation for knowledge-intensive industry, such as Computer Integrated Manufacturing Systems (CIMS) which integrates CAD, CAM, CAPP (Computer-Aided Production Planning) and CAT (Computer-Aided Testing) to accomplish various production tasks, such as taking order, production planning, design, manufacturing, testing, sales and management. This high-performance automation does not allow too much intervention from human experts within the operation process.

The technological advances in automatic control have addressed the research interests of intelligent control, which encompasses the theory and applications of both artificial intelligence (AI) and automatic control (Figure 1.2). Intelligent process control is developed for implementing process automation, improving product quality, enhancing industrial productivity, preventing environmental and hazardous risks, and ensuring operational safety. Two objectives of intelligent control are.

1. enhancing product quality through automation, and
2. improving efficiency through AI techniques.

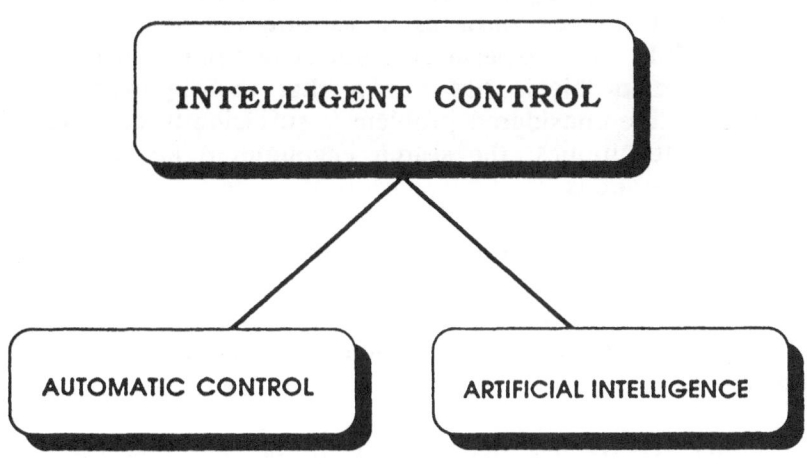

Figure 1.2 Intelligent control

4

Currently, computers have been widely used in engineering applications, but the use has been limited almost exclusively to purely algorithmic solution. In fact, many engineering problems are not amenable to purely algorithmic computation. They are usually ill-structured problems that deal with non-numerical or non-algorithmic information and are suitable for the use of AI techniques, especially expert systems [Weise and Kulikowski, 1984].

As a new advanced technology frontier, AI has been widely applied to various disciplines, including control engineering. This study aims at processing non-numerical information, using heuristics and simulating human being's capability in problem-solving. In fact, a much better terminology for this field was suggested as "complex information processing system", rather than "artificial intelligence". Today, instead of continuing arguing which terminology is better, we concentrate on investigating the use of powerful computation technology in AI research.

An expert system is also a computer program that acquires the knowledge of human experts and applies it to make inference for the user with less training or experience in solving various problems. Expert systems provide programming methodology for solving ill-structured engineering problems which are difficult to be handled by purely algorithmic methods. An expert system is so constructed that it does allow us to capture the way that people reason and think. The experience from developing expert systems for engineering problems has shown that their power is most apparent when the considered problem is sufficiently complex. By means of AI techniques, the search encountered above could be reduced to less options.

For a long time, there has been existing a big gap between new technology and its applications. One of the main reasons is that academia pays little attention to industrial applications. Another reason is the complication of modern science and technology.

Many efforts has been put to investigate and implement real-time intelligent control techniques for industrial applications. There are four main reasons for us to develop intelligent process control systems:

(a) The control problems involved in industrial processes are usually ill-structured, and difficult to be formulated. In such processes, mathematical modelling is not amenable, and purely algorithmic methods are difficult to use. However, AI techniques provide programming methodology for solving these ill-formulated engineering problems.

(b) In industrial manufacturing processes, operating conditions are frequently changed based on different production criteria. There exist many periodic operation procedures. Intelligent systems are suitable for use in such an environment.

(c) The stochastic occurrence of operational faults requires emergency handling in manufacturing processes. Past experience has shown that intelligent fault diagnosis systems are very powerful in dealing with such complex situations.

(d) Industrial process control always deals with the uncertain and fuzzy information. The conventional control systems fail to process such information. However, intelligent systems can process the imprecise information.

(e) Intelligent control is a new technological challenge, which may change the methodology of process control.

In the design of intelligent control system, knowledge processing techniques, including knowledge acquisition, representation, integration, management and utilization, are the main research topics of intelligence engineering which is a new engineering and technology research field.

Traditionally, expert system development mainly relied on the knowledge engineer who, by the definition, was a computer scientist and had the knowledge of artificial intelligence and computer programming. Knowledge engineers interviewed domain experts to acquire the knowledge, then built the expert systems. With the much powerful hardware platforms and more user-friendly programming environment, the increased computation capability of our control engineers, a new era of AI applications is coming to enable engineers to program the expert systems to handle their problems at hand. As we know, during the development of expert systems, knowledge acquisition is the most important but difficult task. Even with the help from knowledge engineers, some private knowledge (such as heuristics and personal experience as well as the role of thumb) may still be difficult to transfer. A new generation of engineers, namely intelligence engineers, are to be trained to meet the needs in this subject. The name of intelligence engineer was set to distinguish the knowledge engineer. An intelligence engineer is a domain engineer, (for instance, a control engineer), who has the certain domain knowledge in the related application area. Through a comparatively short period of training process, she/he learns the basic AI techniques, gets the hand-on experience on programming expert systems. Then, the intelligence engineer could build up much better expert system to solve her/his domain problem, and extend AI applications successfully. Intelligence engineering involves applying artificial intelligence techniques to engineering problems, and investigating artificial intelligence theoretical fundamentals and techniques based on engineering methodology. Figure 1.3 demonstrates the objectives of intelligence engineering. Distinguished from knowledge engineering, intelligence engineering emphasizes on integrating knowledge from different application domains to solve the real world engineering problems.

In the process of building expert systems, acquisition and representation of knowledge are two of the most important steps. The methodology used in this process is different from that in the prototype problem or in the related quantitative simulation program. In the process of developing an expert system, some new methods to solve the problem may be generated, which may complement the solver of the prototype problem [Weise and Kulikowski, 1984; Rao, Jiang and Tsai, 1988]. As usual, the new technique of programming expert systems is sought to guarantee the realization of new algorithms for this problem. This indicates that building an expert system is not just a translation from the existing expertise knowledge into a computer program; it is the production process in which new expertise knowledge can be acquired.

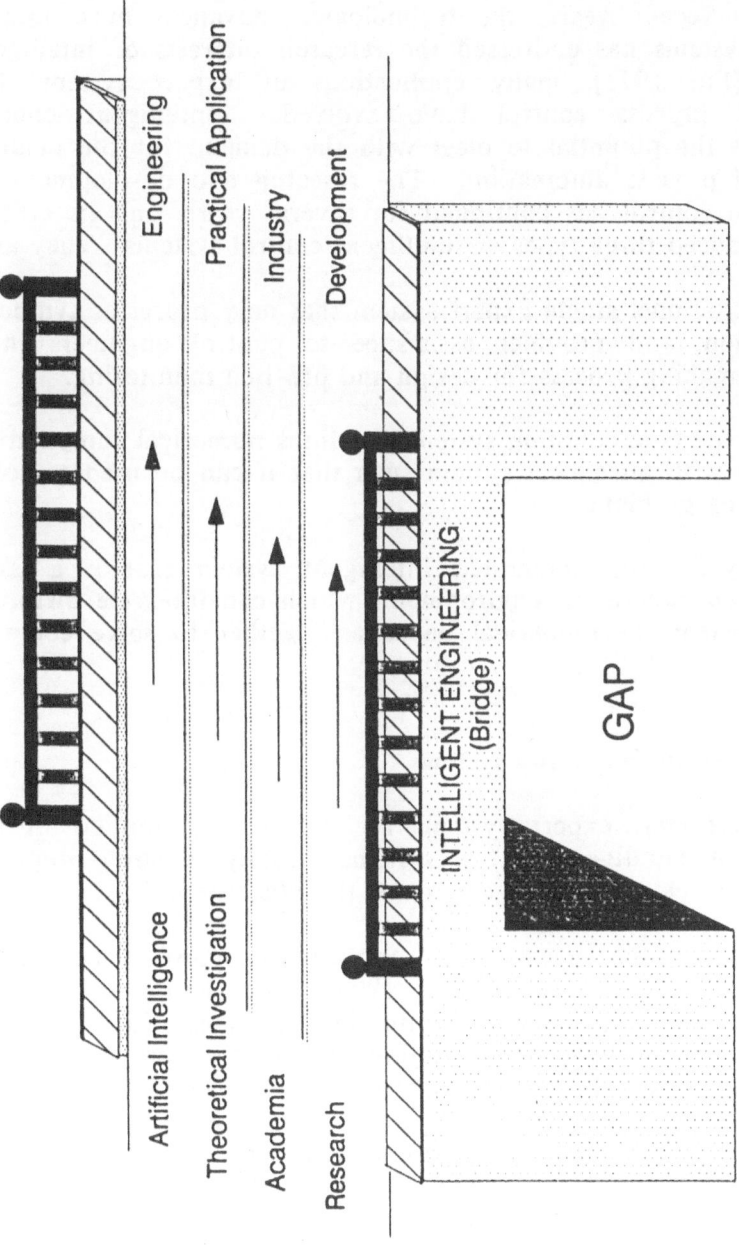

Figure 1.3 Intelligence engineering as a bridge

1.2 Architectures for intelligent control

In recent years, the technological advances in computer control systems has addressed the research interests of intelligent control [Fu, 1971]. many applications of expert systems for advanced process control have evolved. Intelligent control highlights the potential to meet with the demand for the modern industrial process automation. The research and development of intelligent control has continued for several years, and its efforts have produced three types of intelligent control systems. They are:

Type one: single expert system that only processes symbolic information, and provides assistance to control engineers in a decision-making process for design and off-line monitoring.

Type two: coupling system that links numerical computation programs with an expert system such that it can be used to solve engineering problems.

Type three: integrated intelligent system that is a large intelligence integration environment, which can integrate different expert systems or numerical packages together to solve complex problems.

1.2.1 Symbolic reasoning system

Presently, expert systems are extensively applied in the research of intelligent control systems. Many research programs focus on developing intelligent robots [Saridis, 1983].

Among the successful AI applications, most of the expert systems are production systems [Weise and Kulikowski, 1984; Talukdar, et al., 1986]. Production systems facilitate the representation of heuristic reasoning such that expert systems can be built incrementally as the knowledge of expertise increases.

In our projects, OPS5 (Brownston, et al., 1985) serves as one of the programming tools, which is the most widely used language for developing expert systems based on a production system, and consists of three components: a data base (working memory), a knowledge base (production memory) and an inference engine as shown in Figure 1.4. Working memory is a special buffer-like data structure and holds the knowledge that is accessible to the entire system. Each unit of working memory is an attribute-value element. Any attribute that is not assigned a value for a particular instance is given the default value designated as "nil".

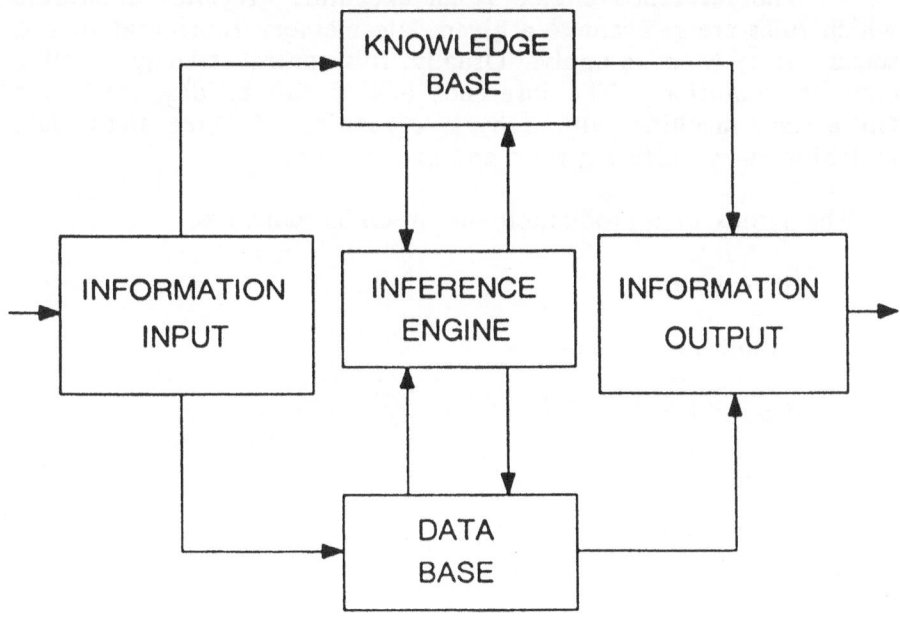

Figure 1.4 Structure of expert system

Production memory contains the general knowledge about the area of the problem. The expertise knowledge for the problem is described by a set of production rules stored in production memory. The typical production rule is described as "IF (condition), THEN (action)." Every production rule consists of a condition-action pair. In OPS5, the condition part is called the LEFT-HAND-SIDE (LHS), and the action part is called the RIGHT-HAND-SIDE (RHS). Each condition element specifies a pattern that is to be matched against working memory. The matching process will be described as we define the syntax of LHS. The action forming RHS are imperative statements that are executed in sequence when the rule is fired. Most of the permissible actions alter the working memory.

The inference engine is an executor. It must determine which rules are relevant to a given data memory configuration and select one of them to apply. Usually, this control strategy is called conflict resolution. The inference engine can be described as a finite-state machine with a cycle consisting of three steps, i.e., matching rules, selecting rules and executing rules.

The syntax of a production rule in OPS5 looks like

(p rulename

condition element

-->

action)

where, "p" represents production rule, LHS and RHS are separated by the symbol "-->".

Figure 1.5 shows the basic stages for developing expert system. The development for the real time intelligent control systems, as my suggestions, should be accomplished by three steps as follows:

The first step: implementing the expert systems for off-line design and training purposes. Since AI techniques don't guarantee perfect solution for problem solving, any prototype intelligent control systems have to be carefully evaluated before using in real time situation.

The second step: using the intelligent control systems that have been very well developed and evaluated in the first step for on-line supervisory control. This step will deal with cooperation with process operators and interface with instrumentation.

The third step: on-line real time intelligent control. However, it still requires to keep human being operator in control loop. In fact, such a feature is a strength for intelligent control, rather than a weakness.

Most of existing expert control systems are implemented for design purposes [Pang and MacFarlane, 1987]. Design is a very complex and not a completely understood process, particularly since it is abstract and requires creativity. Design is largely a trial and error process, in which human design experts often employ heuristics or rules of thumb. An expert system can provide a user-computer interface in such a process to improve the efficiency of the design process [Lamont and Schiller, 1987]. The primary function of the controller can be improved by introducing an expert system into the control system [Astrom, 1989]. The experience from building expert systems also shows that the power of expert systems is most apparent when the considered problem is sufficiently complex [Astrom, 1986]. Moore et al. [Moore, Hawkinson, Knickerbrocher and Churchman, 1984] proposed a real-time expert system for the supervision of a control system, an intelligent fault diagnosis, as well as alarm and performance analysis. Similar investigations were made by Taylor and Frederick [1985] and SRI International [Wright, Green, Fiegl and Cross, 1986]. These expert systems are developed for a single and specific purpose.

Fuzzy set theory was proposed by Zadeh (1978) to represent inherent vagueness. Recently, it has been applied to expert-aided control systems to deal with the imperfect knowledge of control process. For example, a set of fuzzy rules and a fuzzy temporal model are built in the expert system so that the dynamic behavior of the process can be well described [Qian and Lu, 1987].

In the analysis and synthesis of control systems, simulation is a major technique. The traditional simulation techniques are algorithm-based. They are often inflexible and provide limited means to the user. In fact, such techniques can not clearly simulate the dynamic behavior of the controlled processes.

Recently, a knowledge-based simulation system has been suggested to solve the problems encountered above [Zeigler, 1985; Shannon, et al., 1985]. The segregation of the database, knowledge base and inference engine in the expert system allows us to organize the different models and domain expertise efficiently because each of these components can be designed and modified separately.

IDSOC (Intelligent Decisionmaker for the problem-solving Strategy of Optimal Control) is a rule-based simulation system that can handle various complicated decisionmaking problems of optimal control [Rao, Tsai and Jiang, 1988]. Programmed in OPS5 [Brownston et al., 1985], it can modify its knowledge base and deal with uncertainty of input information. The configuration of reasoning at three levels and utilization of "filter rules" make IDSOC potentially powerful. For the time being, IDSOC can only construct the optimal solution structure, while symbolic integration and numerical computation are not available. Expert systems for symbolic integration have been successfully developed [Slagle, 1963; Moses, 1967]. However, no means are provided for them to be used in IDSOC. Lack of numerical computation makes IDSOC unable to operate in the real-time environment.

A functional approach to designing expert simulation systems for control that performs model generations and simulations was proposed by Lirov and his colleague [1988]. They chose the differential games simulator design to build this simulation system. The knowledge representation of the differential games models is described using semantic networks. The model generation methodology is a blend of several problem-solving paradigms, and the hierarchical dynamic goal system construction serves as the basis for model generation. This discrete event approach, based on the geometry of the games, can obtain the solution generally in much shorter time. Cooperation between control systems is achieved through a goal hierarchy. Logic programming, such as Prolog-based tools, seems to be an adequate way for implementation.

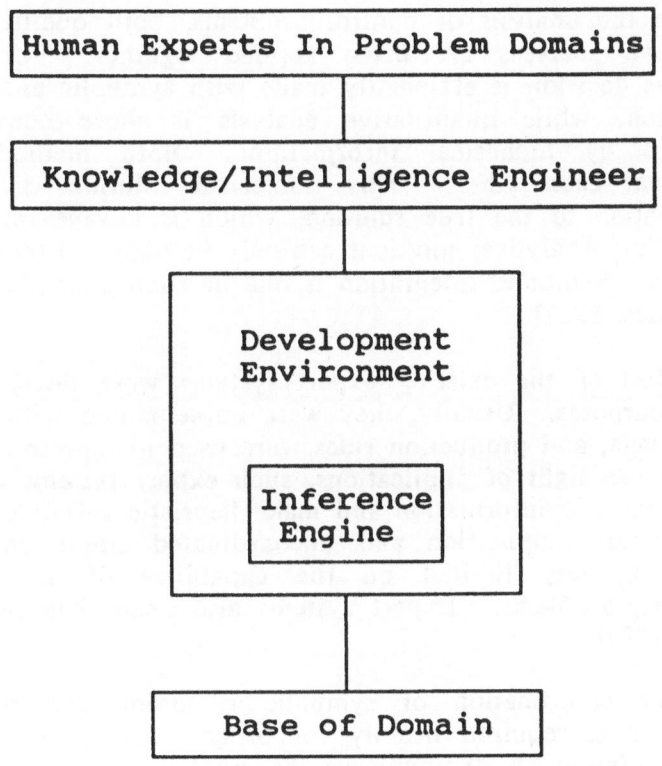

Figure 1.5 Expert system development stages

1.2.2 Coupling system

Historically, process control follows the development of mathematical control theory. Recently, some reports stated that mathematical modeling is not the only means to describe the controlled process [Rao and Jiang, 1988]. The knowledge we know about the world is not well captured by numbers, since our reasoning is not well modeled by arithmetic [Davis, 1987]. Many control problems are ill-structured engineering problems that deal with non-numerical information and non-algorithmic procedure, and are suitable for the application of AI techniques [Buchanan, 1985].

In the analysis of control problems, both qualitative and quantitative analyses are often applied together. Usually, a qualitative decision is efficiently made with symbolic and graphic information, while quantitative analysis is more conveniently performed by numerical information. Both methods often complement each other. Any numerical solution is only an approximation to the true solution, which is always represented analytically. Analytical solutions can only be obtained by symbolic processing. Symbolic integration is one of such examples [Slagle, 1963; Moses, 1967].

Most of the existing expert systems were developed for specific purposes. Usually, they were implemented with a Lisp-like language, and production rules were used to represent domain expertise. In light of applications, such expert systems can only process symbolic information and make heuristic inference. Lack of numerical computation and uncoordinated single application make them very limited on the capability of solving real engineering problems. Expert systems also need data processing [Shirley, 1987].

The coordination of symbolic reasoning and numerical computation is required heavily for process control with expert systems. Figure 1.6 distinguishes the symbolic reasoning system (expert-aided system) with the coupling system from the viewpoint of software architectures. In IDSOC, a set of numerical algorithms to compute certainty factors are coupled in the process of symbolic reasoning [Rao, Tsai and Jiang 1988]. SFPACK [Pang, 1990] incorporates expert system techniques in design package then support more functions to designers. Written in Franz Lisp, CACE-III [James, Frederick, Bonissone and Taylor, 1985] can control the start-up of several numerical routines programmed in FORTRAN. Similar consideration was taken into account by Astrom's group [Astrom, 1986]. ROBEX was written in C language in order to call numerical routines [Lewin and Morari, 1987]. More and more, the coordination of symbolic reasoning and numerical computing in knowledge-based systems attracts much attention. Recently, several methods of integrating expert systems were proposed [Kitzmiller and Kowalik, 1987]. A few developers tried to develop expert systems with conventional languages, such as FORTRAN, so that these expert systems can be used as subroutines in a FORTRAN main program. Others suggested to develop expert systems in conventional languages in order to achieve integration. However, these methods prohibit developing and using the individual program separately. This makes acquiring new programs very difficult, and is not cost-effective. Another disadvantage is that the procedural language environment cannot provide many good features that Lisp provides, such as easy debugging and allowance for interruption by human experts.

Numerical languages often have a procedural flavor, in which the program control is command-driven. They are very inefficient when dealing with processing strings. Symbolic languages are more declarative and data-driven. However, it is very slow for symbolic languages to execute numerical computations. We have realized that if applied separately, neither symbolic reasoning nor numeric computing can successfully address all problems in design and analysis. Complex problems cannot be solved by purely symbolic or numerical techniques [Kitzmiller and Kowalik, 1987; Wong, Dong and Blanks, 1988]. Coupling of symbolic processing and numerical computing is desirable to use numerical and symbolic languages in different portions of a software system.

Currently, not many expert system development tools or environments provide the programming techniques for coupling expert systems applications. For example, it is very difficult to carry out numerical computation in OPS5 [Brownston, Farrell, Kant and Martin, 1985; Rao, Jiang and Tsai, 1988]. However, many artificial intelligence software system developers are now building the general-purpose tools for coupling systems that will be benefit to the expert systems applications to engineering domains.

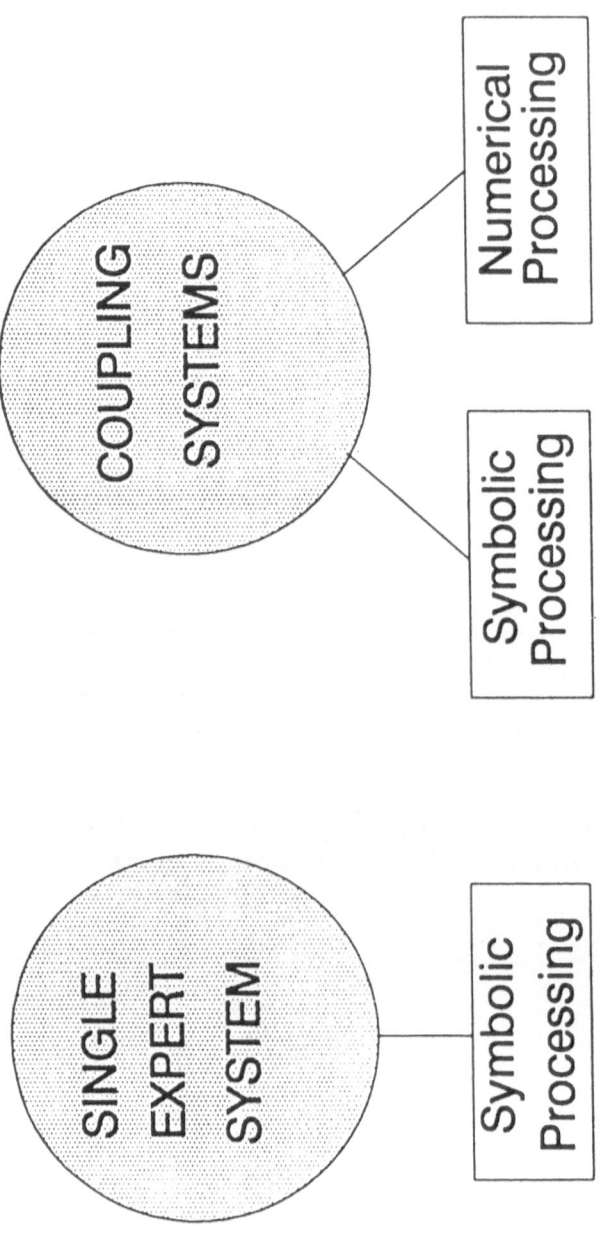

Figure 1.6 Expert system and coupling system

1.2.3 Integrated intelligent system

The need to integrate different numerical computation tools within a package for designing control systems has been realized. On the other hand, the need to integrate intelligence is getting apparent. As mentioned above, the existing expert system can only be used alone for a particular purpose inflexibly. For example, we can not integrate the expert systems that have been available, even though each of them was well developed for a specific task. As we know, the best way to solve the complicated problem by expert systems is to distribute knowledge and to separate domain expertise. In such a case, several expert systems may be used together. Each of them should be developed for solving a sub-domain problem. Here, we face the problem of knowledge integration and management.

Many expert systems can only be used alone for a particular purpose inflexibly. They lack coordination of symbolic reasoning and numerical computation, integration of different expert systems, efficient management of intelligent systems, and capability of dealing with conflict facts and events among the various tasks. In addition, it may be difficult to modify knowledge bases by end users other than the original developers.

YES/L1 [Cruise et. al, 1987] (Yorktown Expert Systems Language One) is intended to exploit the full power of both rule-based programming and traditional languages, which extends the capabilities of the production systems architecture while integrating a number of ideas from traditional programming languages and from specialized expert system shells. The Pilot Associate program [Lizza and Friedlander, 1988] is developed to exploit the new computing technology, such as knowledge-based systems, parallel processing, and speech system, in which the key issue also deals with knowledge integration. Now, one of the remaining bottlenecks for intelligent systems to be applied commercially is to integrate both quantitative and qualitative methods together.

An integrated intelligent system is proposed by Rao, Jiang and Tsai [1987]. It is a large knowledge integration environment, which consists of several symbolic reasoning systems and numerical computation packages. The integrated intelligent system is illustrated by Figure 1.7. These software programs may be written in different languages and be used independently. They are under the control of a supervising intelligent system, namely, a meta-system. The meta-system manages the selection, operation and communication of these programs.

The key issue to construct the integrated intelligent system is to organize a meta-system, which can thus be referred to as a control mechanism of meta-level knowledge [Davis, 1982]. A meta-system has its database, rulebase and inference engine, but it decomposes its activities into separated, strictly ordered, phases of information gathering and processing. The main functions of a meta-system are described as follows:

1. To coordinate all symbolic reasoning systems and numerical computation routines in an integrated intelligent system.

2. To distribute knowledge into separate expert systems and numeric routines so that the knowledge bases of these expert systems are easier to be changed by the commercial users other than their original developers.

3. To acquire new knowledge efficiently.

4. To find a near optimal solution for the conflict solutions and facts among the different expert systems.

5. To provide the possibility of parallel processing in the integrated intelligent system.

6. To communicate with the measuring devices and the final control elements in the control systems and transform various input/output signals into the standard communication signals.

Integrated intelligent system is very demanding to support the integration for different expert systems and numerical computation packages in order to realize the industrial automation in the knowledge intensive stage. This new software integration platform can process different knowledge (analytical and heuristic knowledge), different information (symbolic, numerical and graphic information) as well as different computer languages. It has attracted much research attention, it is the main topics in this manuscript. I will discuss it in details in the following sections.

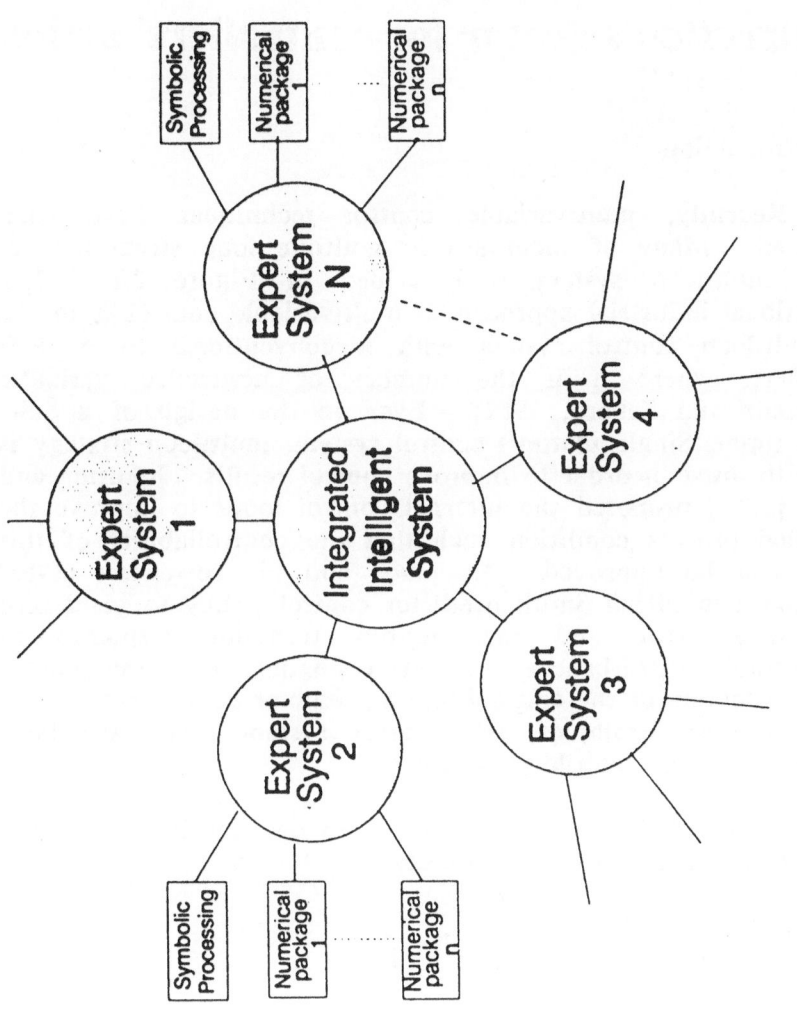

Figure 1.7 Integrated intelligent system

2 Direction selector for controllers' action

2.1 Introduction

Recently, multivariable control techniques have been developed. Many of them are in multiple loop structure. A typical multiloop system is described in Figure 2.1. The conventional industrial approach to multivariable control is to use the multiloop control system with n conventional PI or PID controllers, where n is the number of controlled variables [Gagnepain and Seborg, 1982]. Even in the design of a SISO (Single Input, Single Output) control system, multiloop strategy is frequently used in order to improve control results. Johnston and Barton [1984] proposed the internal control loops to improve the controlled process condition such that the controllability of this system can be improved. Watanabe and his coworkers [1983] suggested a modified Smith predictor control policy to yield zero steady-state error and the desired transient responses to unmeasurable disturbances without changing the input-output transfer function of the original Smith predictor control system. In the above two examples, the control system may have been restructured into a multiloop system.

In designing a control system, a key decision is to design a controller in the following two phases:

1. choose the suitable control law,
2. choose correct controller's action direction.

Usually, the work in phase 1 concentrates on the control theory, but that in phase 2 deals with both the control theory and the instrumentation. So far, among many reports on the multivariable control [Rosenbrock, 1969; MacFarlane, 1972; Owens, 1981], the successful techniques for choosing the direction of the controller's action in a multiloop control system have not yet been developed, even though they are very important for industrial environment. In practice, industrial control engineers are normally concerned with the installation of instrumentation. The designers of process control systems always have to deal with the procedure of choosing the direction of controller's action. This directly affects the process production and safety. The procedures often vary, depending on the production conditions and operation requirements. The traditional method relies on experiments and experience, which are time-consuming and error-prone. Wang [1980] introduced a method for this procedure. Another design approach was suggested by Rao and Jiang [1986] and the software package for that problem has been developed. In view of applications, these methods seem to be somewhat inconvenient for use to design multiloop control systems with a large number of loops.

Our goal is to provide an intelligent program to solve this problem. Along this line, an expert system is developed to aid engineers in choosing the direction of controllers' action in a practical multiloop control system. The expert system, called IDSCA (Intelligent Direction Selector for the Controller's Action), can make heuristic reasoning and intelligent decision. IDSCA is based on production rules, but its ability of performing heuristic reasoning makes it potentially significant for industrial design. Essential to our expert system is a knowledge base of the expertise of chemical process control designers. Based on the expert knowledge, the inference engine of IDSCA can perform intelligent decision and choose the suitable valve type and controller's action direction for a process control system design. Results show that the new system can enhance chemical process production and safety.

22

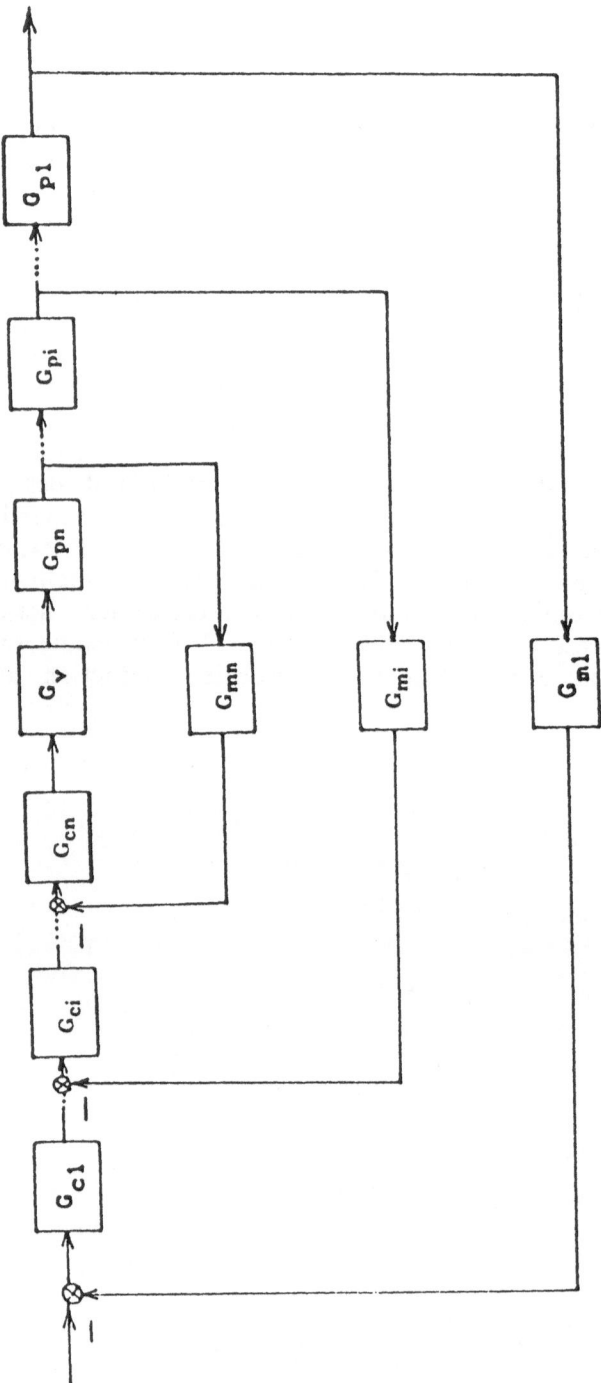

Figure 2.1 Block diagram for multiloop system

2.2 Philosophy of building AI systems

As powerful tools, computers have been widely used in the engineering field, but the use has been limited almost exclusively to purely algorithmic solutions. Many engineering problems are not amenable to purely algorithmic solutions. Usually, to deal with these ill-structured problems, an engineer relies on his own or other experts' judgment and experience. Knowledge-based expert systems provide a programming methodology for solving this kind of ill-structured engineering problems.

The applications of expert systems to control problems attract more attention and the capability for expert systems to deal with difficult and complicated problems becomes more apparent through some successful applications [Astrom, Anton and Arzen 1986; Moore et al., 1984]. An expert system is an intelligent computer program that acquires the knowledge of human expertise and applies it to make heuristic inference for the user in solving various problems. This kind of knowledge-based computer program can provide the powerful capability of helping a human handle the ill-structured problems that are hardly solvable by purely algorithmic methods. It solves the problems at the level of experts in the specific domain.

One reason for us to develop IDSCA is that most of the information and the problem solving method in IDSCA is heuristic. In the process of solving a problem, the information provided by the user is about the input-output relation of the concerned process as well as safety considerations, operation procedures, and so on. The results sent to the user are the decision to choose the action direction for each of the controllers in the solved system and the suggestion about the selection of the valve type. These pieces of non-numerical information can be processed by AI methodology efficiently. Moreover, the strategy used in IDSCA is non-algorithmic. The purpose of IDSCA is to help the user select suitable design models. Every possible selection is based on the inference of related events. Such decision-making can not be described by mathematical formulas or algorithmic expressions. It is, however, suitable for the use of AI techniques [Buchanan, 1985].

Our experience in developing IDSCA has indicated that expert systems should be applied to the problem which is very complicated [Rao, Jiang and Tsai, 1988; Astrom, Anton and Arzen,

1986]. The traditional method for choosing the direction of the controller's action is time-consuming and ineffective for solving the problems with the loop number basic components (process, valve and controller), each of which has two possible options. The options of the 26 possible design models for a single loop system is 2^2. For a n-loop system, the number of options is $(2^2)^n$. When n is large enough (say, n = 10), the number of options can exceed one million! It is not feasible for an engineer to search every possible solution manually or even with a CAD package. With the help of an AI technique, the search options can be largely reduced. Our intelligent system IDSCA can handle the problems for selecting the correct direction of the controller's action for any multiloop control system. Through a user-friendly interface, IDSCA asks the user to provide the necessary information, then performs the heuristic reasoning and provides the solution quickly. High efficiency of IDSCA truly shows the powerful potential for AI systems to deal with complicated problems.

The other experience from developing IDSCA indicates that building an expert system is not just a translation from the existing knowledge into a computer program. It is a process in which the new expertise knowledge may be generated and acquired. The new knowledge may complement the solver of the prototype problem. In the next section, we will introduce the new design criterion that is generated during the process of building IDSCA.

An Adaptive Feedback Testing System (AFTS) is proposed to develop an automatic testing and modifying mechanism for the high performance expert systems. Our first implementation has been applied in IDSCA for this purpose.

The function of AFTS comprises two parts: self-testing and self-modifying. The error-prone solution can be found by self-testing, while the reliable result can be reconstructed by self-modifying. Such an intelligent function is necessary for the applicable intelligent programs. The other feature of IDSCA is that a meta-level control architecture is hierarchically constructed in order to manage the knowledge base effectively. The use of task-independent meta-level knowledge can reduce the search of task-dependent object-level knowledge. These ideas will be illustrated in the subsequent sections.

2.3 Adding new rules while running program

IDSCA is capable of modifying the knowledge base. Production rules in IDSCA can handle most of the related problems. However, in some special cases, the suitable production rules for the particular situation may not exist in the knowledge base to represent the special events. Two methods are available for solving this problem. The first method is to write all production rules in advance for every possible event. This could result in inefficient processing and is memory-consuming. The other method is to create new specific production rules when a special problem is encountered. Needless to say, the latter is a much better approach than the former.

Such a special case occurs when the user can not identify the characteristics of the processes. In this case, new production rules can be created to deal with this specific problem and help the user to identify the process characteristic.

In our expert system, OPS5 serves as the programming tool, which is the most widely used language for developing expert systems based on production systems. By using the action of "build", we can construct new specialized rules. In the OPS5 production system, the syntax of a production rule that generates another new rule looks like:

```
(p rulename
   (condition 1)
   (condition 2)
   ......
   (condition n)
-->
   (action 1)
   (action 2)
   ......
   (bind (new rulename (genatom))
     (build // (new rulename)
     (condition A)
     (condition B)
       ......
     (condition K)
   -->
     (action A)
     (action B)
```

```
......
(action Q)
)
(action i)
)
```

where, "p" represents the production rule, in which condition sequence and action sequence are separated by the symbol "-->". Each rule has a condition-action pair. When the pattern, which contains condition 1, condition 2, condition 3, ..., and condition n, is matched against working memory, a sequence of actions, action 1, action 2, action 3, ..., and action i, will be formed. One of these actions is to create another new production rule in production memory. Similarly, the new rule contains a sequence of condition elements, such as condition A, condition B, ..., and condition K, and a sequence of actions, such as action A, action B, ..., and action Q. Note that in this new rule, the action of "genatom" is to create a symbolic atom, and then the action "bind" assigns the atom to a new generated rule as its name. The action of "build" adds to rule memory the new production rule, which may be applied when the knowledge base is to be modified. A detailed description about OPS5 can be referred to [Brownston, et al, 1985].

2.4 Advisory Configuration

The fundamental requirements of a chemical plant are: operation safety, equipment protection, saving energy, high product quality and low cost. Among them, operation safety and equipment protection are the most important and critical.

During the design of selecting the controller's action direction, the first step is to determine the open-close type of control valve in terms of operation safety and production conditions. The basic criterion to choose the valve type is to assure the security of the process operation and equipment when accidents occur.

As we know, the selection of the valve type is even more complicated than the identification of process characteristics. It deals with the operation procedure, production conditions and safety consideration. Usually, these requirements conflict with one another. In such a case, IDSCA first takes the safety factor into account and performs the heuristic reasoning to seek a suitable solution.

For the convenience of application, IDSCA determines the valve type with two different approaches. When the valve type has been selected by the user, IDSCA immediately assigns the corresponding action direction function to the attribute of the valve block in the inner most loop. In case the user cannot choose the valve type, IDSCA will create new production rules in its knowledge base to help the user select the valve type. Through the menu-driven interface, IDSCA asks the user to provide the following information:

1.   production safety consideration, such as the possible occurrence of accidents as the control signal supplier is shut down or the controller is out of order.

2.   operation procedure and requirements.

3.   production conditions, such as the possible location of valves, the relation between manipulated variable and measured variable, the specific requirement to the valve, and physical limitations.

With the above information at hand, IDSCA then makes inference based on the domain expertise and chooses a plausible valve type as a candidate. This solution is evaluated by the Adaptive Feedback Testing System (which will be described in the next section). If the set of conflict evidences is a comparative minimum, the valve type will be suggested as an optimal choice. With the agreement of the user, the corresponding action direction function is assigned to the valve block.

The configuration of selecting the valve type, for the time being, can only provide assistant and consultant advice, rather than a perfect solution. This feature, as pointed out by Stephanopoulos [1986], is a strength rather than a weakness because it allows the interjection of the human's accumulated knowledge about the specific problem at hand, as the required knowledge may not be available in IDSCA.

2.5 Architecture for meta-Level control

The production memory of IDSCA is composed of several rule bases based on different functions. Not only to set up a hierarchy in structure, but also to manage these production rules, we organize them into two different classes of rules, that is, meta-rules and object-rules [Davis and Lenat, 1982].

The key issue in the design of architecture for IDSCA is to develop a control mechanism of the meta-knowledge for the selection and application of the object-knowledge. The meta-rules provide the means to allow IDSCA to reason about controlling of object-rules. The hierarchical architecture of meta-level control is described in Figure 2.2. This architecture may be represented by a frame as follows:

Table 3   Framework of IDSCA

1. input interface model (RB-1)
2. valve design model (RB-2)
3. controller design model (RB-3)
4. Adaptive Feedback Testing System model
    4.1 distributing model (RB-4.1)
    4.2 testing model (RB-4.2)
    4.3 modifying model (RB-4.3)
    4.4 managing model (RB-4.4)
5. output interface model (RB-5)
6. IDSCA managing model (RB-6)

The structured listing of facts is used to build rule bases. Based on its function, each of the rule bases performs a particular action. RB-1 provides the communication with the user and receives the input information. RB-2 and RB-3 govern the design of the valve and the controllers, respectively. RB-4.1 sends the design results to the testing model or the modifying model. RB-4.2 performs self-testing, while RB-4.3 is used to modify the design if necessary. RB-5 processes the output information and sends the final decision or suggestions to the user. RB-4.4 and RB-6 are the meta-rule bases that govern the selection and application of object-rule bases mentioned above.

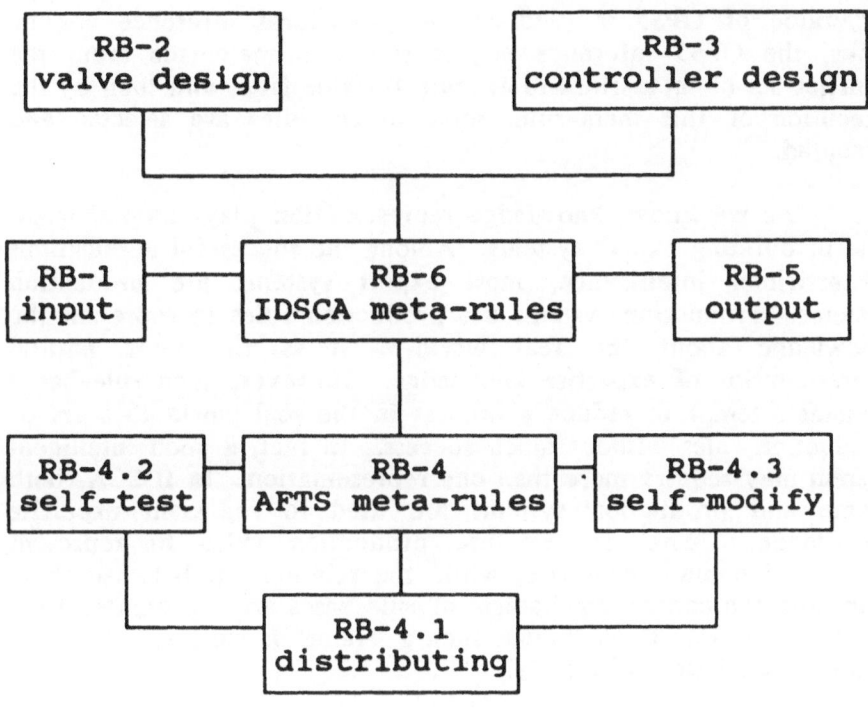

Figure 2.2   Rule base organization of IDSCA

IDSCA works in an interdisciplinary field and its expertise is deep knowledge. Usually, the use of these rule bases is in a certain order. For example, RB-1, RB-2, RB-5 and all the rule bases of AFTS are inferred during the design of the valve type. At first, RB-1 is used to collect and process input information. Then both RB-1 and RB-5 are used to communicate with the user interactively while RB-2 is working. If any conflict is found during the self-testing, RB-2, RB-4.3 and RB-5 will begin to work, and so on. In such a complicated procedure of selecting rules, the most important thing is how to describe knowledge about knowledge, i.e., to know which kinds of knowledge are useful for particular kinds of jobs. Meta-level control strategy allows us to represent knowledge about knowledge. In IDSCA, the sophisticated procedures of selecting and employing rules are governed by two meta-rule bases RB-4.4 and RB-6.   The conflict-set resolution

30

technique of OPS5 is used as the procedural inference engine. Thus, the OPS5 inference engine selects a meta-rule from the conflict-set of meta-rules to execute the rule first, and then by the execution of this meta-rule, some object-rules are selected and executed.

As we know, knowledge representation plays an important role in building expert systems. Among the successful applications of artificial intelligence, most expert systems are production systems. Production systems use production rules to represent the knowledge about the real world. It is the most natural representation of expertise knowledge. However, such rule-based systems attempt to reduce a process in the real world to a set of production rules without much success. In fact, a good intelligent system may require more than one representation. In IDSCA, both frame and production systems are used to represent expertise knowledge. That is, we use production rules to represent individual domain expertise, while the relationship between these rules and the control mechanism of rule bases are represented by a frame structure. Even though such a "frame" is simple, it helps us manage rule bases efficiently.

Like MELD [Thompson and Wojcik, 1984], the object-level rules of IDSCA are task-dependent, while its meta-level rules are task-independent. This architecture may facilitate the search through the object-rule bases by appropriate uses of meta-rule bases and allow us a free hand in employing and managing interdisciplinary and deep knowledge.

2.6 Illustration

A cascade control system, as shown in Figure 2.3, is a simple but important type of control system, which is frequently applied in an industrial environment, particularly in chemical industry. With two different measurements, the cascade control system consists of two control loops: the primary (outer) loop controls the process output variable Y_1, while the secondary (inner) loop controller uses the output of the primary controller as its set point. The secondary loop can be treated as a single block after the direction of its controller's action has been chosen. Then, the direction of the controller's action in the primary loop can be selected.

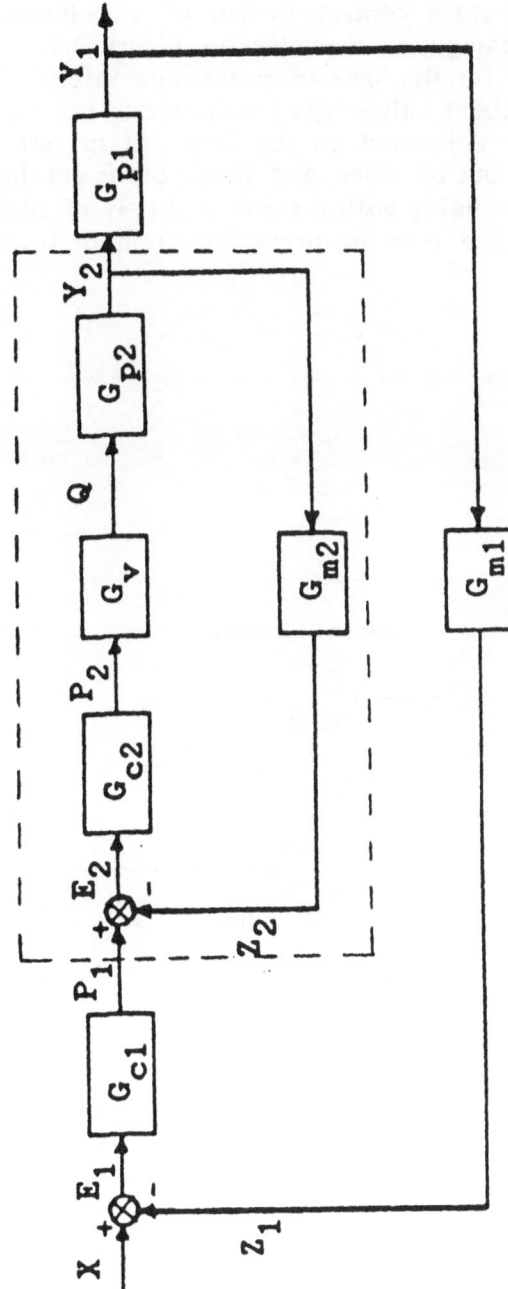

Figure 2.3 Block diagram of cascade control system

Consider a temperature control system of a polymerizer with a cascade control strategy, as described in Figure 2.4. An air-to-close valve is used for the sake of production safety. The reasoning of the advisor about valve-type selection and the testing of AFTS about the valve are based on the fact that in case the controller or equipment is out of order, and air supply is cut down, the valve should open thoroughly with maximum supply of cooling water. The polymerizer can then be prevented from destruction due to overheating.

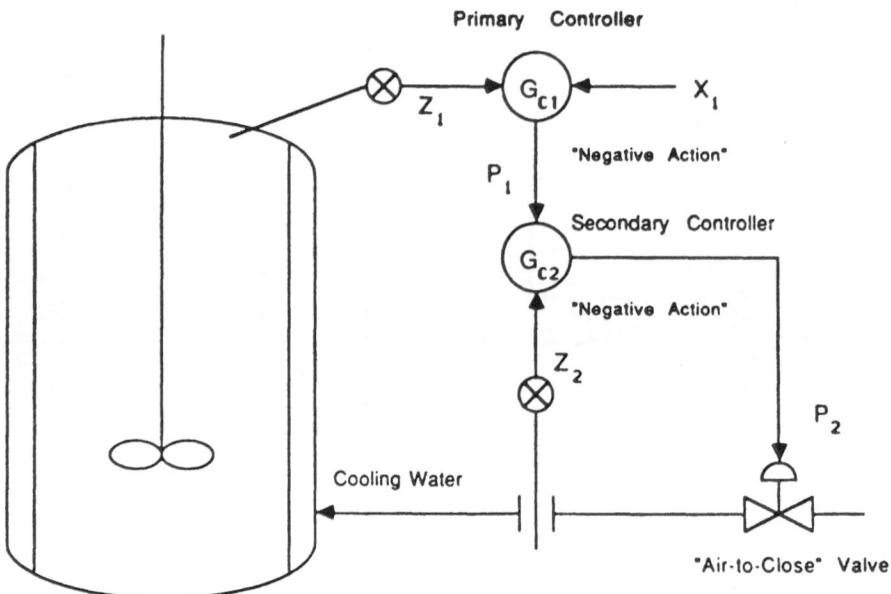

Figure 2.4 Cascade control of a polymerizer

In the secondary loop, process output (flow of cooling water) increases when the process input (valve opening) increases. This suggests that the secondary process characteristic is "positive". For the primary loop, the process characteristic is "negative", since the its output (temperature of the furnace) decreases as its input (flow of cooling water) increases.

The reasoning process of IDSCA is described as follows:

Receive and process input information
Select the valve type
Test the selected valve type by AFTS
Assign "negative" to the action direction function for the valve
Assign "positive" to the action direction function for P_{c2}
Assign "negative" to the action direction function for P_{c1}
Infer the action direction function for secondary loop
Determine the action direction function for G_{c2}
Check the iteration index
Define action direction function for G_{v1} by Equivalence Principle
Infer action direction function for primary loop
Determine the action direction function for primary controller G_{c1}
Test and modify the design results through AFTS
Output the final decision to the user

The reasoning process of AFTS is demonstrated as follows:

Assume a small increase at the output Y_2 of the secondary loop
Infer the change of P_2 based on the direction action of G_{c2}
Infer the change of Q based on the type of the valve G_v
Infer the change of Y_2 based on the characteristic of P_{c2}
Set results as final decision if Y_2 decreases
Modify the results if Y_2 increases
Feed modified results back to AFTS for further testing
Assume a small increase at the output Y_1 of primary loop
Infer the change of P_1 based on the direction of action G_{c1}
Infer the change of P_2 based on the direction action of G_{c2}
Infer the change of Q based on the type of the valve G_v
Infer the change of Y_2 based on the characteristic of P_{c2}
Infer the change of Y_1 based on the characteristic of P_{c1}
Set results as final decision if Y_1 decreases
Modify the results if Y_1 increases
Feed modified results to AFTS for further testing

As the final decision, IDSCA prints the results on screen:

DESIGN RESULTS

CHOOSE AIR-TO-CLOSE VALVE FOR G_v (control valve)
CHOOSE NEGATIVE ACTION FOR G_{c2} (flow-rate controller)
CHOOSE NEGATIVE ACTION FOR G_{c1} (temperature controller)

2.7 Summary

An expert system for selecting the direction of the controller's action and the type of control valve in a multiloop system is developed. IDSCA can not only perform heuristic reasoning but also test and modify its results adaptively. It paves a way for our research on the use of intelligent systems for computer-aided control system design. IDSCA seems to be relatively simple but effective for the design of industrial process control systems. The important significance behind the practical application is the investigation of both new design criterion and Adaptive Feedback Testing System. The former may complement the knowledge of process control, while the later will stimulate the development of intelligent systems with high reliability and performance. In short, such a process of building intelligent systems is of benefit not only to AI research, but also to the development of control theory. Finally, the reader should note that the methodology provided in this paper is based on the block diagram for the control system. For the instrumentation in which the definition of the error signal is reversed, the results for the direction of the controller's action are just opposite.

3 Integrated intelligent system

3.1 Review and background

The technological advances in automatic control have addressed the research interests of intelligent control, which encompasses the theory and applications of both artificial intelligence (AI) and automatic control (AC). Many successful developments have been reported in recent years [James, 1987; Lemont and Schiller, 1987; Rao and Jiang, 1989]. The primary function of a controller can be improved by introducing an expert system into a control system [Astrom, Anton and Arzen, 1986]. Moore et al. [1984] proposed a real-time expert system for the supervision of control systems, and intelligent fault diagnosis, as well as alarm and performance analysis. Similar investigations were made by James, Taylor and Frederick [1985] and SRI International [Wright et al., 1986]. Many research programs in intelligent control focus mainly on developing intelligent robots [Saridis and Valavanis, 1988]. Along the way, the architectures and environments for intelligent control have been raised as key issues [Rao, Jiang and Tsai, 1988; Handelman and Stengel, 1987; Ionescu and Trif, 1988].

As the part of our ongoing work on intelligent control, several expert systems have been developed. IDSOC (Intelligent Decisionmaker for the problem-solving Strategy of Optimal Control) is a rule-based system that can handle various complicated decisionmaking problems of optimal control [Rao, Tsai and Jiang, 1988]. Programmed in OPS5, it possesses the capability to modify its knowledge base and to deal with information uncertainty. The configuration of reasoning at three levels and the utilization of "filter rules" make IDSOC potentially powerful. Its weakness is that while the optimal solution structure can be constructed, symbolic integration and numerical computation are not yet available in IDSOC. Expert systems for symbolic integration have been developed successfully [Slagle, 1963; Moses, 1967]. Several research projects on using symbolic manipulation in designing control systems have been reported [Akrif and Blankenship, 1987; Eldeib and Tsai, 1988], although no means are provided for them to be used with IDSOC. Lack of numerical computation makes IDSOC unable to operate in the real-time environment.

IDSCA (Intelligent Direction Selector for the Controller's Action) was developed to assist in designing the controller in multiloop control systems [Rao, Jiang and Tsai, 1988]. It assists control engineers in selecting valve types and to direct the controller's action. Also written in OPS5, IDSCA provides some new features, such as its modifiable knowledge base, an Adaptive Feedback Testing System that provides the means to enhance the reliability of design results, and the hierarchy of meta-level control strategy. As we know, in designing a control system, the key issue is to design a controller is to choose the suitable control law, and to choose the correct direction of the controllers' action.

IDSCA accomplishes the objective in phase 2 successfully. This indicates that it can be used in various computer-aided control system design projects. However, a major problem is how to link IDSCA with other existing design packages.

Having reviewed the existing expert systems, we find that most of them were developed for specific purposes. Many were implemented with a Lisp-like language, and production rules were used to represent domain expertise. In light of applications, such expert systems can only process symbolic information and make heuristic inferences. A lack of numerical computation and uncoordinated single applications limit their capability to solve real engineering problems.

3.2 Quantitative and qualitative analyses

As we know, computers are mainly used in numerical computation for engineering applications. In fact, many engineering problems are not amenable to purely algorithmic computation. They are usually ill-formulated problems, which deal with non-numerical or non-algorithmic information and are suitable candidates for the use of intelligent system techniques.

Expert systems are developed to solve such ill-formulated problems which are difficult to handle by purely algorithmic methods. An expert system is also a computer program that acquires the knowledge of human experts and applies it to make inferences for the user with less training or experience in solving various problems. The knowledge base of an expert system contains two kinds of knowledge: the public knowledge and the private knowledge (expertise). It is a specific computer program, implemented by a specific programming technique, and used to solve a specific problem in a specific domain. The segregation of the database, the knowledge base and the inference engine in the expert system allows us to organize the different models and domain expertise efficiently because each of these components can be designed and modified separately.

We often use qualitative and quantitative analyses together in the problem-solving of engineering problems (Figure 3.1). Usually, qualitative decisions are mainly based on symbolic and graphical information, while quantitative analysis is more conveniently performed using numerical information. Each method often complements the other.

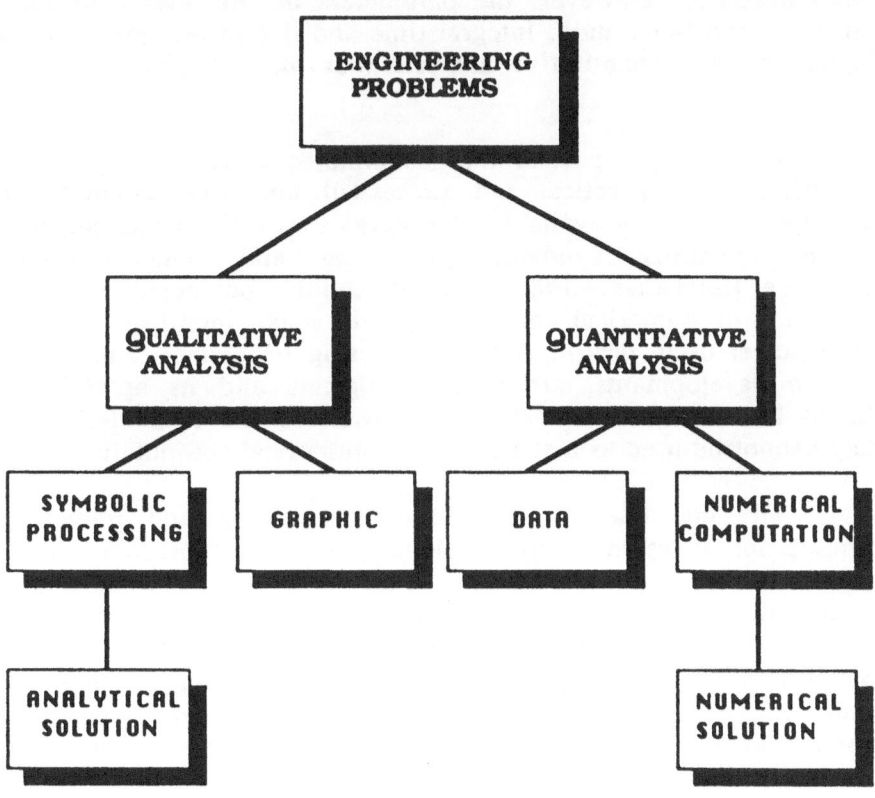

Figure 3.1 Qualitative and quantitative analyses

For any numerical solution, whatever it is how perfect, it is always an approximation to the true solution. True solution is always represented analytically. Analytical solutions can only be obtained by symbolic processing.

However, a main disadvantage of existing expert systems is their inability to handle numerical computation. This makes these expert systems less useful for many engineering problems. In the real-time control environment, we require not only a qualitative description of system behavior, but also a quantitative analysis. The former can predict the trend of the change of operating states, while the latter may provide us with a means to identify the change range of these states. For example, qualitative analysis of a temperature control process may suggest the use of a three mode PID controller. However, the parameters of this PID controller, namely proportional gain, integral time and derivative time, mainly depend on the quantitative analysis that may require numerical computations.

Moreover, as part of the accumulated knowledge of human expertise, many practical and successful numerical computation packages are already available. We agree that artificial intelligence should emphasize symbolic processing and non-algorithmic inference [Buchanan, 1985], but it should be noted that the utilization of numerical computation will make intelligent systems more powerful in dealing with engineering problems. Like many modern developments, artificial intelligence and its applications should be viewed as a welcome addition to the technology, but they cannot be used as a substitute for numerical computation.

The coordination of symbolic reasoning and numerical computation is essential to developing intelligent process control systems for industrial applications. More and more, the importance of coordinating symbolic reasoning and numerical computing in knowledge-based systems is being recognized. It has been realized that if applied separately, neither symbolic reasoning nor numeric computing can successfully address all problems in engineering design and analysis. Complicated problems cannot be solved by purely symbolic or numerical techniques [Jacobstein, Kitzmiller and Kowalik, 1988; Wong, Dong and Blanks, 1988].

Integrating different numerical computation tools within a package for designing control systems has been realized [Colgren, 1988]. The need to integrate intelligence is also becoming apparent. Therefore, many existing expert systems can only be used alone for a particular purpose, and inflexibly. For example, we cannot integrate CACE-III [James, et al. 1985] with IDSCA [Rao, Jiang and Tsai, 1988], neither can we combine SAINT [Slagle, 1963] or CONDENS [Akrif and Blakenship, 1987] with IDSOC [Rao, Tsai and Jiang, 1988], even though each of these expert systems was well developed for its specific task.

However, we are now trying to solve more complicated problems with the methodology of expert systems. The best way to solve a complicated problem by expert systems is to distribute knowledge and to separate domain expertise. In such a case, several expert systems may be used together, each of them having been developed for solving a sub-domain problem. Here, we face the problem of knowledge integration and management.

As we know, many existing expert systems are developed with specific techniques [Bulter, Hodil and Richardson, 1988]. Their knowledge bases can only be modified by either the original developers, or someone who clearly understands the software structure and design details of the expert systems. In light of commercial applications, the maintenance and modification of expert systems is the key to the development of high-performance intelligent programs that can solve more complicated problems. Current expert systems are particularly short of this capability.

Very recently, several methods of integrating expert systems have been proposed. A few developers tried to develop expert systems with conventional languages, such as FORTRAN, so that these expert systems could be used as subroutines in a FORTRAN main program.   Others suggested embedding expert systems in conventional languages in order to achieve integration. However, these methods prohibit us from developing and using the individual programs separately. This makes acquiring new programs very difficult.   Another disadvantage is that the procedural language environment cannot provide many good features that Lisp supplies, such as easy debugging and allowance for interruption by human experts. It is clear that one of the bottlenecks for the commercial application of intelligent systems is how to integrate and manage knowledge efficiently.

3.3 Meta-system and its main functions

So far, many expert systems have been developed. However, their capability to deal with engineering problems is very limited.

The following summarizes the main disadvantages of the existing
expert systems:

1. lack of the coordination of symbolic reasoning and
    numerical computation,

2. difficulty in modifying knowledge bases by end-users other
    than the original developers,

3. lack of integration of different expert systems,

4. lack of the efficient management of intelligent systems,

5. lack of capability of dealing with conflicting facts and
    events among various tasks.

With increased experience of building expert systems, we
have realized that integrating intelligent systems into a large
environment is often necessary but difficult. We propose a new
architecture to control and manage the large-scale intelligent
systems, which includes the following phases:

1. integration of knowledge of different disciplinary domains;

2. integration of empirical expertise and analytical knowledge;

3. integration of various objectives, such as research and
    development, system design and implementation, process
    operation and control;

4. integration of different symbolic processing systems (expert
    systems);

5   integration of different numerical computation packages;

6. integration of symbolic processing systems and numerical
    computation systems;

7. integration of different information, such as symbolic,
    numerical and graphic information.

Phases 1 and 2 are at the knowledge level. Phase 1 also
indicates the characteristics of modern engineering techniques. In
our case, it means the integration of artificial intelligence and
automatic control, as well as chemical processes. Phase 3 deals

with both the knowledge and the functional levels. Phases 4 through 7 perform their integration function at the functional level, through the problem-solving level to the program level. In what follows, we focus our attention on the latter three phases.

The industrial manufacturing process includes two stages: manufacturing and testing. These two stages are operated based on the information provided from the design stage. There are two flows in the process: material flow and information flow. The material flow is all necessary working stations where workpiece is processed, as well as the transportation of the workpiece between these stations. The information flow can be divided into two branches based on their functions: one is the domain information flow (or technical information flow), another is the meta-information flow (or decision-making information flow). The domain information flow is the processing and transforming of the information about product design, manufacturing and testing in the production system. Th meta-information flow is the processing and transforming of the information about system integration, management and coordination among all stages. The meta-information flow controls the domain information flow and material flow, enables them to combine together and go through the system smoothly. There three flows compensate each other and coexist to form a complete system. In such a way, the integration environment for distributed intelligent system is needed.

An integrated intelligent system is a large knowledge integration environment, which consists of several symbolic reasoning systems and numerical computation packages. These software programs may be written in different languages, and may be used independently. They are under the control of a supervising intelligent system, namely, the meta-system. The meta-system manages the selection, operation and communication of these programs. We discuss it in detail below. For the sake of convenience, a frame representation technique is applied to describe the hierarchical architecture of the integrated intelligent system. The hierarchical structure of the software environment is demonstrated in Figure 3.2.

By integrating intelligent systems, we can achieve efficiency as well as conceptual and structural advantages. This new conceptual design framework can serve as a universal configuration to develop high-performance intelligent systems for many complicated applications.

42

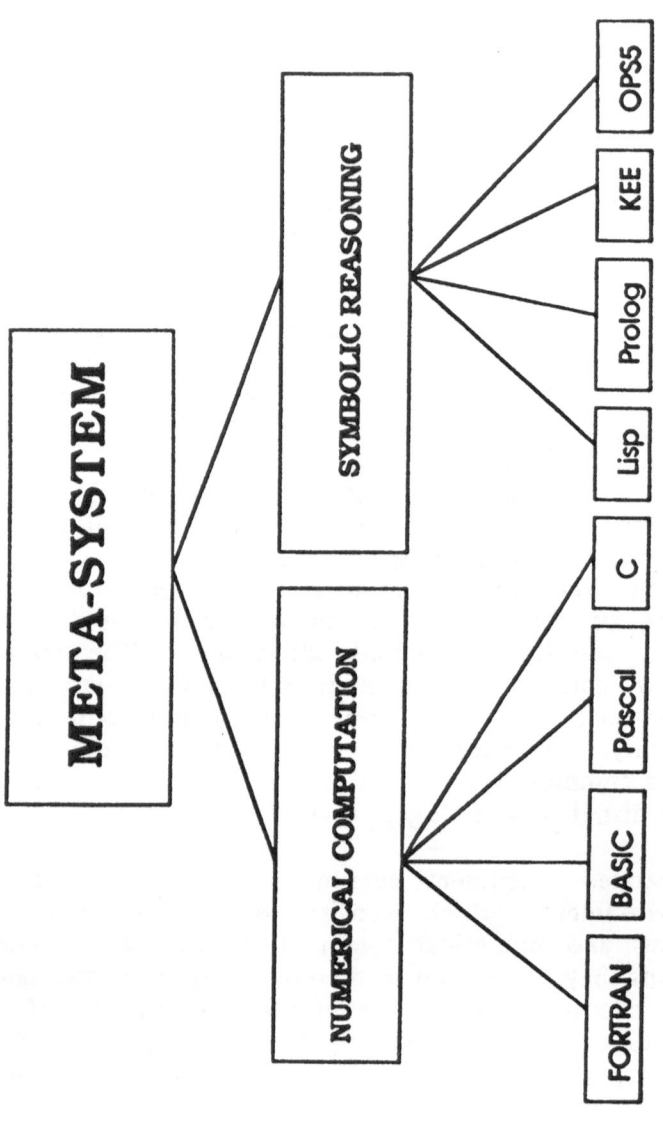

Figure 3.2 Hierarchy of integrated intelligent system

The key issue in constructing the integrated intelligent system is to organize a meta-system. A meta-system can thus be referred to as a "control mechanism of meta-level knowledge", which is defined as "the knowledge about knowledge" by Davis [1982]. Knowledge can be categorized into "domain knowledge" and "meta-knowledge". The former is usually defined as facts, laws, formulae, heuristics and rules in a particular domain of knowledge about the specific problems, whereas the latter is defined as the knowledge about domain knowledge and can be used to manage, control and utilize domain knowledge. Meta-knowledge possesses diversity, covering broader area in contents and varying considerably in nature. Also, it has a fuzzy property of more uncertain than fact. Hence, the representation and knowledge base organization of meta-knowledge should have their own characteristics.

Presently, the concept of meta-level knowledge is a very broad one. However, its application concentrates on implementation of meta-rules that have been successfully used to control the selection and application of object-rules or rulebase [Davis and Lenat, 1982; Rao, Jiang and Tsai, 1988; Orelup and Cohen, 1988]. As yet, we cannot find any reported meta-level techniques which can solve the problems encountered in a large-scale intelligent system.

In software engineering field, the terminology meta system is also widely used. For example, a meta system for software specification environments are also developed [Dedourek, et al., 1989]. These meta systems are developed to support the production of information processing systems throughout their life cycle. They greatly help us to reduce the time and cost of software development, to maintain and improve existing software environments. Needless to say, these meta systems have benefited a lot to database systems and software environments. In this paper, I would like to present another meta system that is developed to handle integration problems in distributed intelligence systems. Such a meta system is different from those discussed by Dedourek and his colleagues [1989] in both concepts and implementations.

Not just extending the concept of meta-level knowledge, but rather proposing an innovative idea, we suggest the development of a new expert system, that is, a meta-system, to control an integrated intelligent system. Like the common expert systems, the meta-system has its database, knowledge-base and inference engine, but it distributes its activities into the separated, strictly ordered, phases of information gathering and processing.

Briefly, the main tenets of my view about the meta-system are:

1. The meta-system is the coordinator to manage all symbolic reasoning systems and numeric computation routines in an integrated intelligent system.

The hierarchy of our integrated intelligent system as described in Figure 3.2 indicates that different types of information (symbolic, graphic and numerical) are utilized and processed together. Even in the branch of symbolic processing, many different languages and tools (such as Lisp, Prolog, OPS5, KEE, EMYCIN, and so on) may be applied to build individual symbolic reasoning systems. In each language family, there also exist different expert systems. For example, among the OPS5 family, there are IDSOC, IDSCA, et. al [Rao, Tsai and Jiang, 1988; Rao, Jiang and Tsai, 1989]. The meta-system controls the selection and operation of all programs in the integrated intelligent system, and executes the translation between different data or programs. For example, it may translate a set of numerical data into some symbolic sentences, and then send the symbolic information to an expert system that is to be invoked next.

2. The meta-system distributes knowledge into separate expert systems and numeric routines so that the integrated intelligent system can be managed effectively. Such a modularity makes the knowledge bases of these expert systems easier to be changed by commercial users other than their original developers.

Expanding the investigation of CACE-III [James, et al., 1985], we propose a distributed architecture to offer flexibility and efficiency in the decision-making process. With the meta-system, the knowledge sources can be separated into individual expert systems and numeric programs, which can be developed and applied for specific purposes at different times. Such a configuration of distributed knowledge allows us to write, debug and modify each program separately so that the overall integrated intelligent system can be managed efficiently. Another primary function is that we can reduce the scope of rule search such that the running time of the overall system is minimized, to accomplish the real-time objective. This function can also help reduce the rule or system interaction.

The basic approach is to divide the domain knowledge for a complex problem into a group of expert systems and numeric routines and then to attempt to limit and specify the information flow among these programs. This approach makes the integrated intelligent environment much easier to be modified.

3. Meta-system is the integrator which can help us easily acquire new knowledge.

The meta-system provides us with a free hand in integrating and utilizing new knowledge. Each of the programs in an integrated intelligent system is separated from every other, and only ordered strictly by the meta-system. Any communication between two programs must rely on translation by the meta-system. Such a configuration enables us to add or delete programs much more easily. When a new expert system or numerical package is to be integrated into the integrated intelligent system, we just have to modify the interface and the rule-base of the meta-system, while other programs are still kept unchanged. The most significant fact is that successfully developed software (symbolic reasoning systems or numerical computation packages) can be applied wherever needed. The time and money required in doing "repeated work" can be saved to make the commercial application of intelligent systems feasible.

4. The meta-system can provide a near optimal solution for conflicting solutions and facts among different expert systems.

Knowledge sources are distributed into various distinctive programs that correspond either to different tasks or to different procedures. However, conflict usually arises because different domain expert systems could make different decisions based on different criteria, even though the data that fire the rules are nearly the same. In particular, chemical process control is the science and technology of automation in chemical processes. Interdisciplinary in nature, it allows the application of the knowledge from electrical engineering, system science and computer science to be extensively applied to chemical manufacturing processes. The intelligent control system for chemical manufacturing processes works in such an interdisciplinary field, and often deals with conflicting requirements, in which various domain expertise is utilized. These expert systems may produce conflict decisions even based on the same information. For example, when a disturbance occurs, the "operation expert" will change the operation state, but the "control expert" may wish to keep the operation state unchanged. In this case, the meta-system can pick an optimal solution from among the conflicting facts when the requirements from different domains contradict each other. This function will play an even more important role in knowledge integration and management in the future.

Currently, a method has been developed to search a near optimal solution in a conflicting decision making process, which is based on the priority ranking, that is, different objectives (such as criteria, facts, and methods) are assigned different priority factors, thus each solution will result in an overall priority factors is similar to that in certainty factor calculation. However, this method greatly relies on the expertise of the person who ranks the priority among, various objectives. It is mot easy to use in general cases.

5. Meta-system provides the possibility of parallel processing in the integrated intelligent system.

In a production system, all rules and data are effectively scanned in parallel at each step to determine which rules are fireable on which data. This inherently parallel operation can perform very well in the integrated intelligent system [Okuno and Gupta, 1988].

The one-to-one connection between a program and the meta-system allows the execution of two or more individual programs at the same time. It is due to the parallel computer that several inference engines are able to deal with simultaneous bidirectional resolution [Habdelman and Stengel, 1987]. For example, forward-chaining can be employed on one processor and backward-chaining on another. This allows dynamic resolution strategies [Burg, et al., 1985]. The execution of a rule in the meta-system may result in several different actions that can invoke two or more programs to operate separately at the same time. Such a function can greatly enhance the quality of real-time control. Also, it may be applied to other computational processes.

6. The meta-system can communicate with the measuring devices and the final control elements in the control systems

and transform various non-standard input/output signals into the standard communication signals.

A real-time control environment always deals with the transformation of communication signals, which are either inputs or outputs, such as electronic (current, voltage, resistance), pneumatic, acoustic or even optical signals on different scales. The meta-system may provide a user-friendly interface to the control systems so that these signals can be exchanged, depending on the needs of the user. For example, let us say that the output water temperature of a boiler rises, and it results in an increased electronic current from a thermocouple. Through the interface transformation of meta-system, the signal may be expressed by some numerical value or symbolic information and be used for computing or reasoning. The corresponding signal will be produced once a final result is obtained in an integrated intelligent system, and can be used to manipulate final control element, such as a control valve.

In order to illustrate the function of a meta-system, let us visualize an integrated intelligent system as a business company. The meta-system acts as a manager who assigns different jobs to the employees according to their expertise, and monitors the progress of their work, as indicated by functions 1 and 2. When a newly-hired employee joins the company, the manager is responsible for providing the new job assignment, and possibly for modifying the work plan. Meanwhile, other people still do essentially the same jobs as they did before. This simulates function 3. If the suggestions from two employees are in conflict, the manager must decide on an optimal solution between them, similar to function 4 of the meta-system. Function 5 can be easily visualized as a situation where the manager can let two employees work on their own projects at the same time but share the same tools and resources. Of course, the function of parallel processing is more sophisticated than the simple management system described here. It requires further investigation and study. The transformation between production plan and market information demonstrates function 6.

4 Implementations of meta-system

4.1 Introduction

The development of new expert systems is changing rapidly both in ease of construction and time required due to expert systems building tools.   In the selection of expert systems developing tools, commercial tools was put as the first priority, thus implemented the first meta-system in OPS5 environment. The reason is very straightforward since a sophisticated, well-endowed development and run-time environment can save us a great deal of work generating basic facilities, including reporting, debugging, graphics database, statistical packages, and other specialized functions. On the other hand, the commercial expert systems building tool industry is still in its infancy [Bowerman and Glover, 1988].   Especially, the commercial tool for building integrated intelligent environment has not been reported at this moment. To face this challenge, other two new meta-systems are implemented in C language environment and TURBO-PROLOG language.   These efforts will pave the way for the future implementation of the meta-systems. I first present the layout of meta-system, then discuss these implementation methods using OPS5 and C in this section, and briefly discuss TURBO-PROLOG implementation issue in Chapter 8 "Gear integrated manufacturing system".

4.2 Meta-system layout

The meta-system layout includes the following six main components:

1. Interface to external environment

The interface to external environment builds the communication between the users and internal software systems as well as among the external software systems.   The interface includes an icon structured menu that consists of windows, data structure, security module, editor module.   Windows display different media of information. Data structure is used to receive information.   Security module identifies the classes of users. Editor module helps users access all knowledge bases and databases at meta level and subsystem level. Furthermore, the interface will provide more intelligent functionalities such as natural language recognition for oral inputs and computer vision technique for handwriting inputs. The interface will play a key role in an open structured software system in two ways: the first one is to codify human expertise into the computer system such that it can adopt the most creative intelligence and knowledge in decisionmaking; another is to communicate with other intelligent software systems to extend the system into a much larger scale for the more complicated tasks.

2. Meta-knowledge base

Meta-knowledge base is the intelligence resource of the meta-system. It serves as the foundation for meta-system to carry out the managerial tasks. The meta-knowledge base consists of a compiler and a structured frame knowledge representation facility. The compiler converts the external knowledge representation that is obtained through the editor and is easy to be understood by users into internal representation form available to the inference mechanism for processing. The structure of knowledge representation can be production rule or frame or a combination of both. Characterized with diversity and variety in nature, the meta-knowledge may be better represented in Object-Oriented frame structures. There may be several modules in the frame for different components of meta-knowledge used for different purposes. For example, the communication standardization module for heterogeneous subsystems and the conflict resolution module and others can be formed for general management purpose, whereas the module of knowledge about subsystems and the task assignment module and others must be built according to each of specific problems. The advantage of open structured meta-knowledge base allows intelligent functionalities to enter the meta-knowledge base to engage more duties in decision making, which was solved by human experts before.

The database in the meta-system functions as a global database for the integrated intelligent system, distinguished with those database in the subsystems which only attach to the individual subsystems they belong to. The database contains an editor, an interface to the inference engine, a management system and a physical storage structure. The interface converts the external data representation form into an internal form. The control of data flow in the database is provided either by the inference engine, depending on the corresponding module in the meta-knowledge base, or by users at the certain security classes. The management system will carry out data processing function.

4. Inference mechanism

Due to the diversity of the meta-knowledge and the variety of its representation form, the inference mechanism in the meta-system adopts various inference methods, such as forward chaining, backward chaining, exact reasoning, inexact reasoning, and soon. The inference mechanism conducts operation and processing on the meta-knowledge. Additionally, it also carries out various actions according to the results from reasoning, e.g., passing data between any two subsystems, and storing new data in the database. Therefore, there are some functional modules in the mechanism, which further extends the functionality of the inference mechanism.

5. Static blackboard

The static blackboard is an external memory for temporary storage of information needed to keep when the system is running. Limited by the on-board memory space, the integrated intelligent system is unable to execute at the same time. In fact, it is unnecessary to run the entire system simultaneously. Very often, the meta-system and all subsystems are run on the distributed hardware environment so that there must be a buffer area in the external memory for any two subsystems to exchange information. Besides data storage, the conversion of data in heterogeneous languages into exchangeable standard form will also be completed on the static blackboard.

6. Interface to internal subsystems

This component of the meta-system is established based on each specific application. The interface connects any individual subsystems which are used in problem solving and under the control and management by the meta-system. Each module of the interface converts a nonstandard data form from a specific subsystem into a standard form in the integrated intelligence environment. The conversion among the standard forms of different languages is carried out by the meta-system.

4.3 Implementation in OPS5 environment

The meta-system was first being implemented with OPS5 on a VAX 11/780 computer, running the Unix operating system. OPS5 is a Lisp-based tool [Brownston, et al., 1986]. In the Lisp language, there is no difference between numerical data and symbolic attributes. A message can thus be an arbitrary Lisp expression. Lisp programming environment supports good debugging features. OPS5 was developed at the Carnegie-Mellon University, and is the most popular tool to be used to develop production rule-based systems. Various versions of the OPS5 expert system development languages are now available for many different computers. For example, OPS5+, for the IBM PC, the Macintosh, and the Apollo Workstation, is available from Computer Thought Company, Plano, TX. OPS5 supports a forward chaining inference process. Its pattern matching methodology permits variable bindings. However, it does not provide facilities for sophisticated object representation, and has difficulty with numerical computation. It is not an easy tool for the nonprogrammer to use.

In order to meet the requirements for integrated intelligent system, we have to make some modifications to OPS5. The main tasks here are to increase numerical computation power to the extent that it is capable of solving general engineering problems, and to enable OPS5 to initialize a sub-process (i.e., to enable another expert system to work), and to terminate its sub-processes, as well as to constantly monitor the progress of its sub-processes.

OPS5 does not possess enough numerical operations suitable for an engineering environment. Actually, the commonly used programming language to deal with numerical computation in control systems is FORTRAN. However, OPS5 is built on the top of Lisp. Most Lisp functions can be used in the OPS5 environment through the "external function" declaration. In Lisp, a compiled FORTRAN function can be called through a special loader. By a simple interface, we can call a FORTRAN subroutine into the OPS5 environment. This procedure is shown in Figure 4.1. Besides FORTRAN, C and PASCAL can also be utilized in this manner. Of course, certain disciplines have to be obeyed in order to use those foreign functions. However, to combine FORTRAN code into the OPS5 environment is a great challenge, since any FORTRAN subroutine can only be called by a running FORTRAN main program, and this restricts the use of FORTRAN subroutines within Lisp functions. Fortunately, the configuration of the integrated intelligent system allows us to run FORTRAN subroutines as separate processes, and then to build communication between FORTRAN and Lisp through data files and the interface rule base of the meta-system. Through Lisp external functions, the meta-system that is running under the OPS5 environment can call a numerical computation routine into the integrated intelligent system.

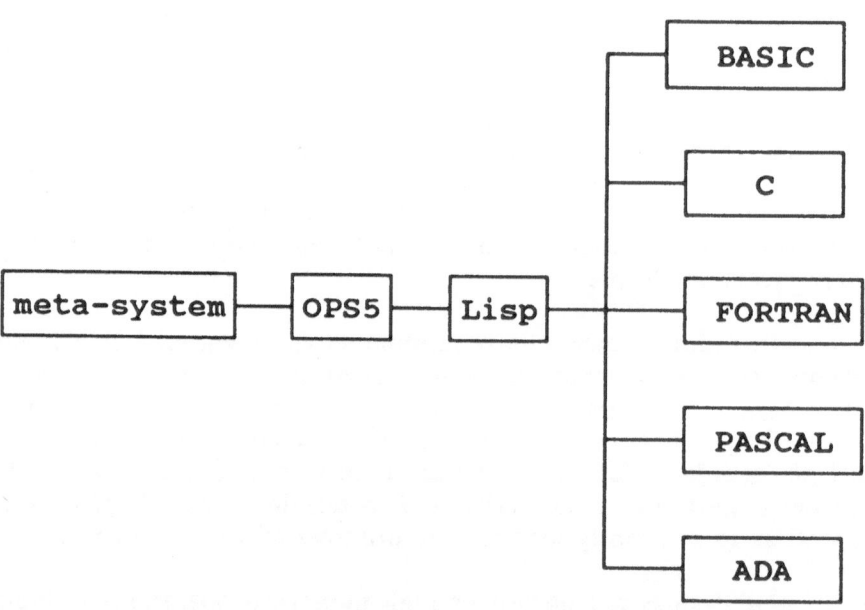

Figure 4.1 Calling heterogeneous langauge numerical routines

        Once the meta-system chooses an expert system (or expert systems) to solve a problem, the meta-system should be able to initialize this expert system (sub-process), to constantly monitor its progress, and to terminate this sub-process whenever necessary. This can be achieved through Lisp commands, such as FORK and WAIT. On the other hand, the communication between the meta-system and expert systems can be made by allocating and maintaining a common area as one-directional pipes, which are monitored by the meta-system and controlled by the meta-rules.

Because all expert systems are dealing with highly interrelated problems, it is convenient for these expert systems to share the same working memory and conflict sets. It is easier to handle a homogeneous environment than a heterogeneous one, since the former has the same data representation and the same inference mechanism. The main tasks here are concurrence control and recency control. The concurrence control problems of homogeneous distributed expert systems resemble those encountered in a distributed database management system, and are crucial to the conflict-resolution strategy in OPS5. Time tags should be issued by the meta-system to every expert system, and care should be taken by the meta-system to decide the recency of each instant generated by different expert systems.

When dealing with heterogeneous distributed expert systems, e.g., when integrating into an integrated intelligent system, the expert systems that are implemented in other environments or tools (such as KEE or EMYCIN), the main issue is how to map or translate different data representations and different inference mechanisms. The meta-system can possess the knowledge of its working expert systems, including the particular expert system's knowledge representation techniques, inference mechanisms, and control strategies. A better way may be to create an expert-systems-capability-knowledge-base, like the UNIX system that stores information about most of the terminals' capabilities in the database TERMCAP. It is only through this knowledge base that the meta-system can communicate with different expert systems and choose the right one to perform a certain task.

4.4 Implementation in C environment

Currently, another meta-system is being implemented with C language environment in my research laboratory. There are four main reasons to use C as an implementation language:

First of all, the C language is versatile in both numerical computation and symbolic manipulation. Its capability to handle numerical operation is much more powerful than Lisp, Prolog, OPS5 and other expert system development tools. It is also superior to FORTRAN and BASIC in term of symbolic processing operation. This advantage makes C easier to integrate different forms of knowledge. Secondly, the C language possesses merits of both high level and low level languages such that it is very flexible and convenient for program coding and control on hardware, especially on the UNIX operating system that is developed in C. Thirdly, the C language can easily access other language environments by interfaces written in mixture of C and the assembly language. Finally, C++ is an object-oriented programming language, and is an extension of C.

The advantage of implementing such a meta-system is that the symbolic process can follow the progress of the numerical process by receiving posted information from the numerical procedure at several steps during number crunching execution. The symbolic process then has the option of letting the numerical procedure continue, of changing some parameters, or of aborting the procedure all together. This contracts favorably with shallow-coupled processes in which the heuristic process invokes a numerical routine via a procedure call, supplies the necessary input information, and passively awaits for the numerical process to finish execution and provide the required output. Such a system configuration is demonstrated by Figure 4.2.

This new meta-system, namely Meta-C, is an expert system building tool that facilitates the deep-coupled integration reasoning and algorithm-based numerical processes, which takes C as the fundamental knowledge representation language and serves as a functional supplement to the C language. The important features of Meta-C are:

(1)   flexible reasoning,

(2)   a high-level representation language, which incorporates extended C data structures to define knowledge bases explicitly, and

(3)   the collaboration of different types of programs by supplying a flexible interface based on a news posting and delivering model.

Since Meta-C knowledge sources are compiled into standard C data structures, it has reasoning and knowledge representation facilities comparable to those of existing tools while at the same time, maintaining the real-time performance offered by low-level C.   Moreover, to ease data communication between a symbol manipulating heuristic process and a data-crunching numerical algorithm, the working memory is unified into a normal C data structure that is partitioned.

A knowledge source defines the purview of the partitioned working memory for it to have access, the triggering preconditions for it to fire, the actions for it to take, and the initializing attributes before the knowledge source is used as well as the working team it belongs to, and so forth. Several related knowledge sources can be grouped together and a "team head," the planner, stipulates how the knowledge sources interact and in what order.

The news posting and delivering model springs from the blackboard control architecture [Hayes-Routh, 1985] and acts as the mediator between heuristic reasoning and the algorithmic numerical procedures, thus serving as the synchronization mechanism between the two processes. Rather than allowing each knowledge source to search the blackboard for the information it requires, the strategy followed is to pass the newly received information through a fast pattern matching scheme and subsequently alter the knowledge sources that may have use for such information. This lowers execution time and favors our goal of real-time performance. The pattern matching network is constructed by means of a compiler, after the knowledge sources are developed.

56

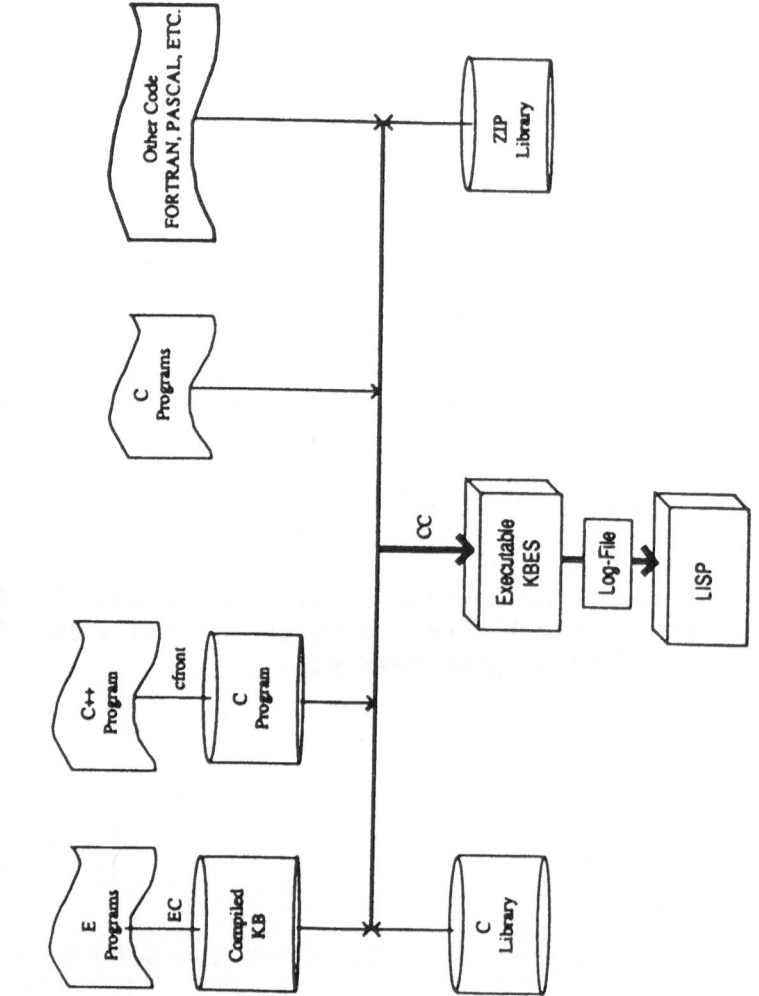

Figure 4.2 Information processing schematics of Meta-C

5 Intelligent optimal control system

5.1 Review and background

Optimal control takes an important place among control engineering and system science. In the past two decades, a lot of research and implementation on optimal control have been reported, but the real applications of optimal control are few. Besides the strict requirement for the controlled system, other obvious obstacles are the complexity of the problem and time-consuming computations.

With the fast development of artificial intelligence techniques, many researchers try to solve decision and control problems by utilizing AI technique. Recently, many applications of expert systems to scientific and engineering problems have been successfully developed, and the capability for expert system to handle the difficult and complicated problems gets more apparent through these successful applications. Rao, Jiang and Tsai (1987) investigated the applications of intelligent systems in chemical Process control. A real-time expert system for the supervision of control system and intelligent fault diagnosis, as well as alarm and performance analysis was proposed (Moore, et al., 1984). Astrom and his colleagues (1986) developed an expert system based on the simple algorithm of PID controller to illustrate that the powerful control function can be obtained by combining convenient control algorithms with an expert system.

5.2 AI approach to optimal control

In general, we first consider an optimal problem described by the following equations:

$$(dx/dt) = A(t)x(t) + B(t)u(t) \qquad (5.2.1)$$

$$x(t_0) = x_0 \qquad (5.2.2)$$

$$x(t_f) = x_f \qquad (5.2.3)$$

We wish to find the variable u that minimizes the cost function J:

$$J(u) = \theta[x(t),t] + \int g[x(t),u(t),t]dt \qquad (5.2.4)$$

subject to

initial constraint: $M(x_0,t_0)=0$ (5.2.5)

terminal constraint: $N(x_f,t_f)=0$ (5.2.6)

This is the well-known Bolza problem which is one of the most complicated forms of a constrained optimization problem and is frequently applied in the optimal control of dynamic systems. For optimal control, most of the constrained optimization problems can be represented by the Bolza problem (Sage and White, 1977). As we know, the procedure of solving a Bolza problem consists of three phases:

(1)   To determine the solution structure by means of solving Euler-Lagrange Equation.
(2)   To obtain the transversality conditions by solving the equations associated with various boundary conditions and constraints on time variables and state variables.
(3)   To construct an optimal solution.

The tasks performed in each of the above processes include:

(a)   decision of problem solving strategy,
(b)   symbolic integration, and
(c)   numerical computation.

Many practical and successful packages for the numerical computation are available. Symbolic integration had been a challenging topic in the pioneer period of AI research. In fact, a symbolic integration system, called SAINT, has been developed (Slagle, 1963). Since the decisionmaking of problem-solving strategy is very critical and complicated, people usually fail to choose the correct strategy such that the suitable solution structure and transversality conditions can not be acquired successfully. In this paper, we will concentrate on the new methodology for problem-solving of optimal control using AI.

For the optimal control problem, the objective of task (a) is to find:

(1)   solution structure,

(2)   transversality conditions,
(3)   suitable technique to solve the related problem,
(4)   the types of Riccati Equations for the linear control problems.

Table 1   Case study of a simple example

variable	symbol possibility	
initial time	T_O unspecified,specified	2
terminal time	T_f unspecified,specified (infinite, $T_f < T_c$, $T_c < T_f < 3T_c$)	4
initial state	X_{Oi} (i=1,2) specified unspecified	4
terminal state	X_{fi} (i=1,2) unspecified, specified (related to X_{Oi} T_f or both, independent)	25
constraint	M_i with M_i, without M_i	4
manifold	N_i with N_i, without N_i	4
control variable	U_i with or without constraint on U_i	4
problem	mathematical form	3
structure	performance index	6

     In order to illustrate the complexity of task (a), we use an example that is a simple problem in optimal control. For convenience, let us consider the state variable x(t) to be a second-order vector. The basic input information and the possibility of their effects on task (a) are demonstrated by Table 1. For this particular problem, the accumulated possibilities of different output results are more than 150,000 ! In fact, this case is not as complicated as many others. For example, we have ignored the situations in which x_{f1} is related to x_{O2}, or both x_{O1} and x_{O2}, and similarly for x_{f2}. If the system is of three dimensions, the corresponding possibilities can exceed 10,000,000 ! Obviously, it will be nearly impossible for the human being to search every promising case for an optimal solution.

With the help of artificial intelligence techniques, the search would be reduced to less options. For this particular problem, our expert system IDSOC can finish the search for all cases and produce the decisionmaking results in a few seconds after the input information has been provided.

Another reason for us to focus on task (a) is that our emphasis is on solving the problem using AI methodology. As pointed out by Buchanan (1985), "AI is distinguished from other area of computing in its attention to both symbolic (nonnumerical) information and heuristic (nonalgorithmic) methods for solving problems. While it certainly does not ignore mathematical expressions of knowledge and variables with numerical values, AI has a special place in computer science in dealing with symbolic inference." It is based on this viewpoint that many AI applications focus on problem solving.

We now consider the problem of selecting the continuously differentiable function vector $x:[t_0,t_f] \to R$, where t_0 is initial time, t_f is terminal time, and R is Euclidean space. For the general form of Bolza problem [Equation (5.2.4)], we introduce a scale function, namely Hamiltonian, as defined in the following form:

$$H[x(t),u(t),\lambda(t),t]=g[x(t),u(t),t]$$

$$+\lambda^T(t)f(x,u,t) \qquad (5.2.7)$$

Using method of Lagrange multipliers, we adjoin the system differential equality constraint, initial constraint manifold, terminal constraint manifold by means of Lagrange multipliers $\lambda(t)$, n, and m and rewrite the cost function as follows:

$$J(u)=\theta[x(t),t]+\xi^T M[x(t_0),t]+\nu^T N[x(t_f),t_f]$$

$$+\int [H(x,u,\lambda,t)-\lambda^T(t)](dx/dt) \qquad (5.2.8)$$

To minimize cost function J(u), the necessary condition is that the first variation in J(u) vanishes. Corresponding control variable u which makes the first variation in J(u) vanish, is the optimal control variable u .

We can form the first variation in Equation (5.2.8) and rearrange it into the following form:

$$\delta J(u) = \int \{ [[(\partial H/\partial x) + (\delta \lambda/dt)]^T \partial x + [(\partial H/\partial k) - (dx/dt)]^T$$

$$+ (\partial H/\partial u)^T \partial u \} dt$$

$$+ \partial t_0 \{ -H - (\partial h/\partial t_0) + (\partial M/\partial t_0)^T \xi \}$$

$$+ \partial t_f \{ H + (\partial h/\partial t_f) + (\partial N/\partial t_f)^T \nu \}$$

$$+ \partial^T x(t_0) \{ \lambda(t_0) - (\partial h/\partial x(t_0) + [\partial M/\partial x(t_0)]^T \xi \}$$

$$+ \partial^T x(t_f) \{ -\lambda(t_f) + [\partial \theta/\partial x(t_f)] + [\partial N/\partial x(t_f)]^T \nu \}$$

$$(5.2.9)$$

To set Eq. (5.2.9) equal to zero so that the necessary condition for a minimum can be obtained, both the integration part and the transversality condition (TC) part must be equal to zero. The integrand in the integration part leads to the well-known Euler-Lagrange equation. The TC part represents terms outside the integration symbol and, when solved, provides the solution of optimal trajectory. The algorithm of setting the integration part and the TC part to zero, respectively, is illustrated by Figure 5.1.

In the top-down algorithm as shown in Figure 5.1, when the integration part is set to zero, the Euler-Lagrange equation for the Hamiltonian are formed, which determine the structure of the optimal solution. If there exists the constraint on the control variable u, ZH/Zu may not be zero. In this case, the maximum principle [Sage and White, 1977] should be used.

For convenience, we define

$$T_0\_term = -H - (\partial \theta/\partial t_0) + (\partial M/\partial t_0)^T \xi$$

$$T_f\_term = H + (\partial \theta/\partial t_f) + (\partial N/\partial t_f)^T \nu$$

$$X_0\_term = \lambda(t_0) - [\partial \theta/\partial x(t_0)] + [\partial M/\partial x(t_0)]^T \xi$$

$$X_f\_term = -\lambda(t_f) + [\partial \theta/\partial x(t_f)] + [\partial N/\partial x(t_f)]^T \nu$$

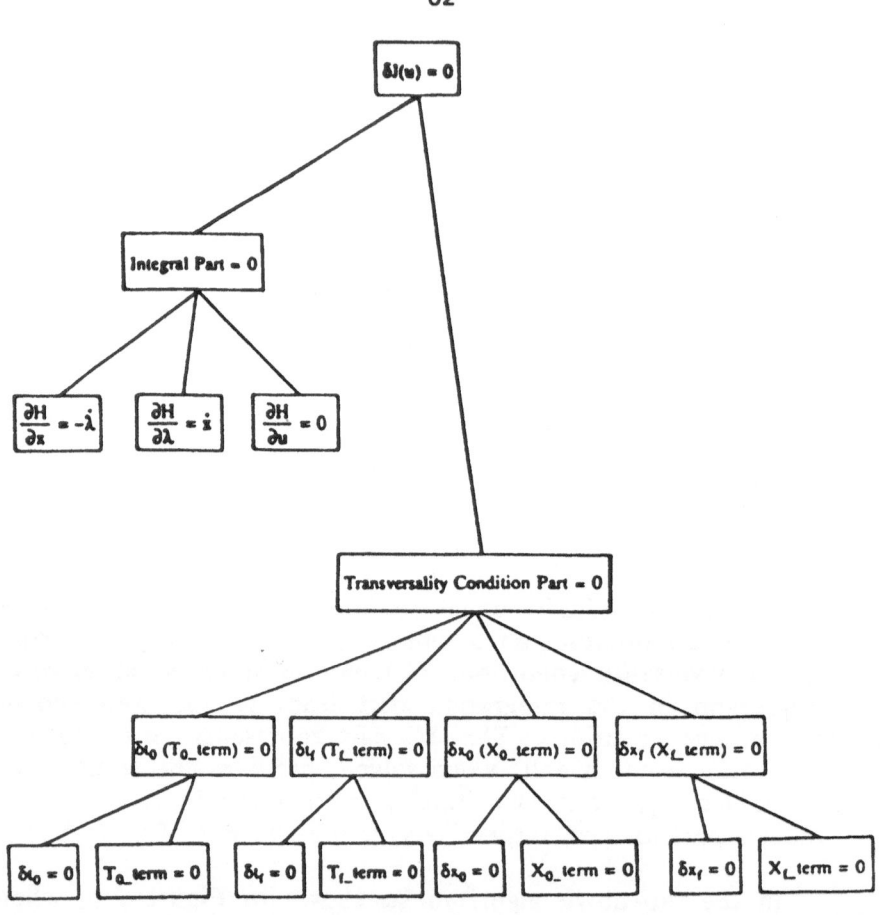

Figure 5.1   Philosophy of problem-solving

63

Then the transversality conditions are given by the following equations:

$$\partial t_0 \, (T_0\_term) = 0 \qquad (5.2.10)$$

$$\partial t_f \, (T_f\_term) = 0 \qquad (5.2.11)$$

$$\partial^T x_0 \, (X_0\_term) = 0 \qquad (5.2.12)$$

$$\partial^T x_f \, (X_f\_term) = 0 \qquad (5.2.13)$$

Note that the T_0_term and T_f_term are scalars, while X_0_term and the X_f_term are vectors.

Generally speaking, if a time variable (t_0 or t_f) or state variable (x_0 or x_f) is specified, then its first variation will be equal to zero. The associated variable_term may be ignored. For example, if t_0 is specified, say t_0=3 seconds, the first variation in t_0 is zero and Eq. (5.2.10) is automatically satisfied. The necessary condition for the first variation of J to be zero at t=t_0 is still guaranteed. When a variable is unspecified, its first variation is in general not equal to zero, and the corresponding variable_term has to be assigned as zero so that an optimal solution can be guaranteed.

The Figure 5.2 demonstrates the relationship between mathematic form and performance form. The detailed strategy for a simple problem is described by Figure 5.3. Here we consider only the case in which the variables and constraints are scalars. In Figure 5.3, ST represents subtree. Figure 5.3 (a) describes the first level of reasoning, the successors of which are the parents of nodes in Figure 5.3 (b). Similarly, the starting nodes of reasoning at the third level are the successors at the second reasoning level.

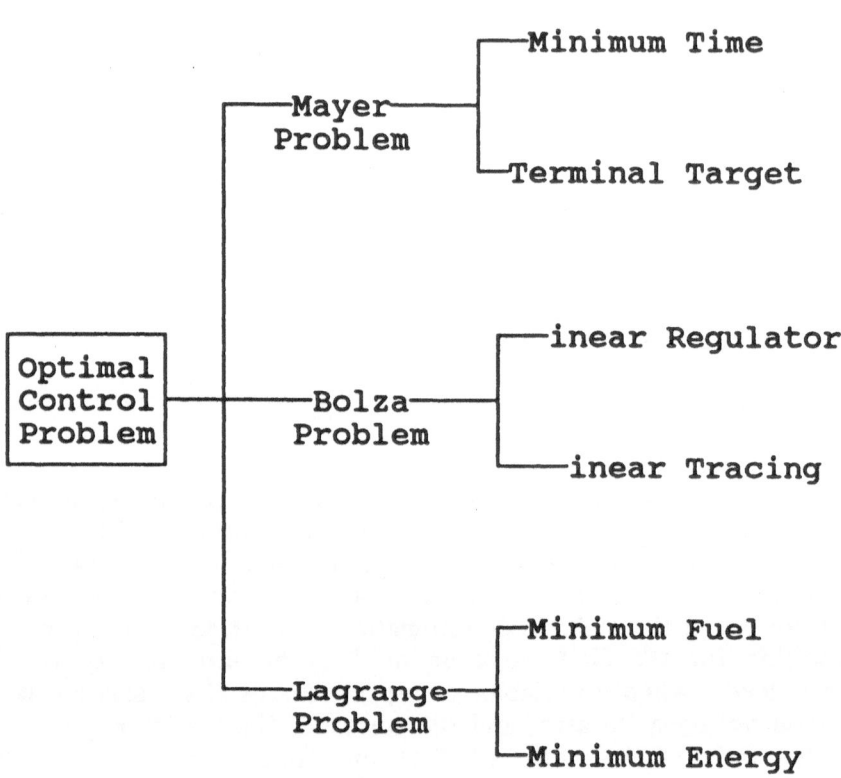

Figure 5.2 Relation between problem classifications

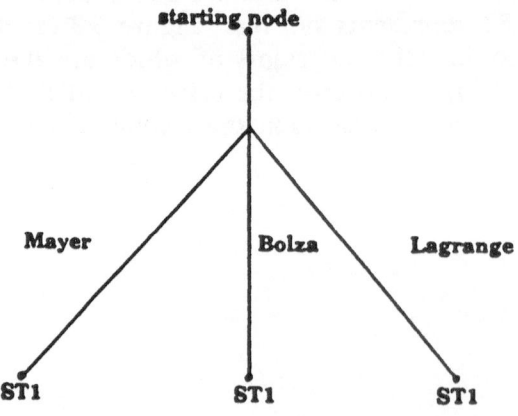

Figure 5.3 (a) Search in the first level

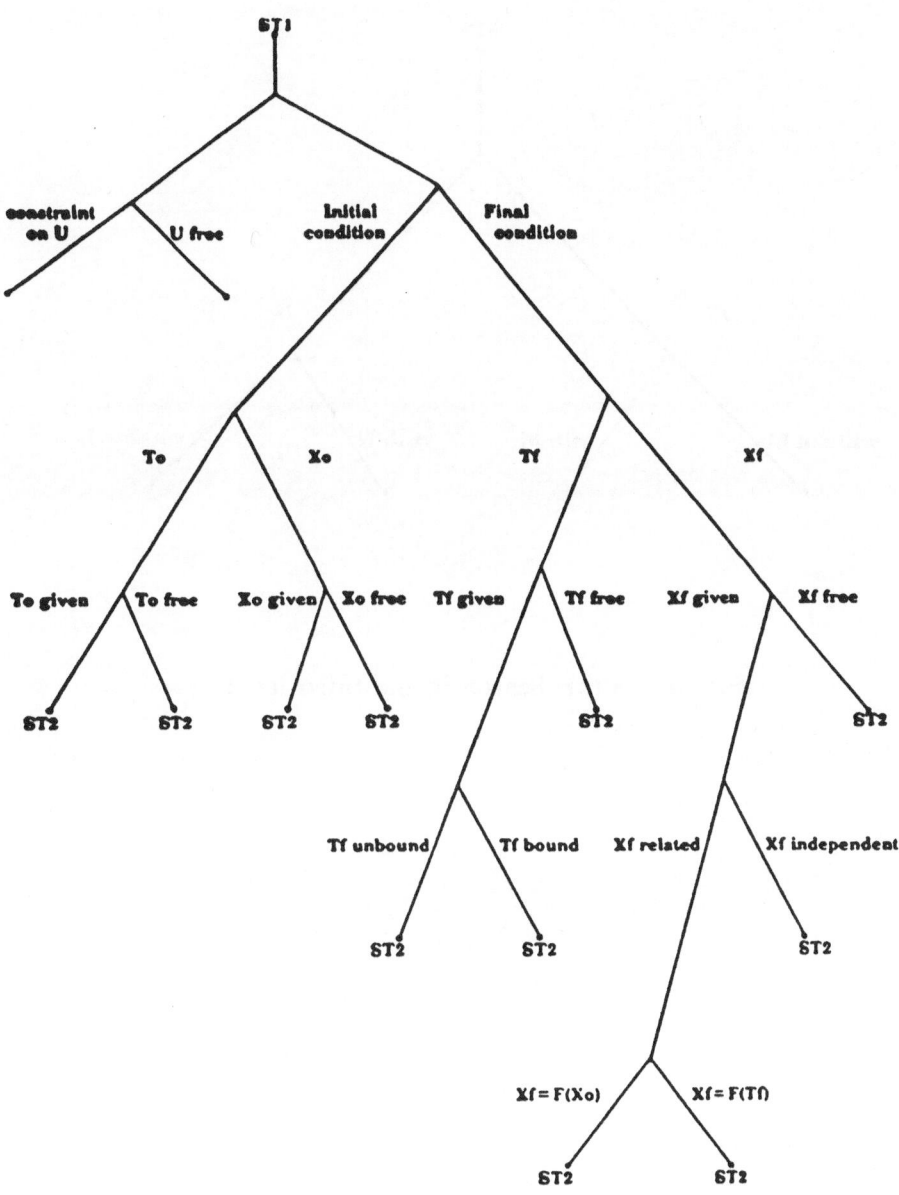

Figure 5.3 (b) Search in the second level

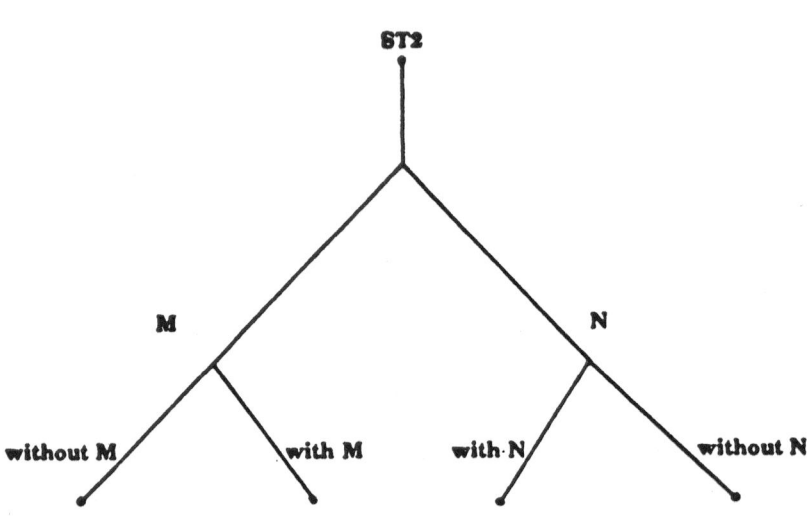

Figure 5.3 (c) Search in the third level

5.3 Knowledge representation

In IDSOC, OPS5 (Brownston, et al., 1985) serves as the programming tool, which is the most widely used language for developing expert systems based on a production system, and consists of three components: a working memory, a production memory and an inference engine. Working memory is a special buffer-like data structure and holds the knowledge that is accessible to the entire system. Each unit of working memory is an attribute-value element. Any attribute that is not assigned a value for a particular instance is given the default value designated as "nil". Production memory contains the general knowledge about the area of the problem. The expertise knowledge for the problem is described by a set of production rules stored in production memory. The inference engine is an executor. It must determine which rules are relevant to a given data memory configuration and select one of them to apply. As an example, a production rule in IDSOC is given as follows:

```
(p Boundary-Condition-To-2
 (Boundary_Condition  ^variable To   ^value unspecified)
  (Constraint_Manifold  ^manifold M   ^structure erased)
 {<a>  (Transversality_Condition
      ^variable_term To_term    ^state original
      ^structure diff_M_over_To-diff_Theta_over_To-H)}
-->
 (modify <a>   ^state modified
  ^structure -diff_Theta_over_To-H)
 )
```

In LHS, the value of the time variable To is unspecified, and the initial constraint manifold M is erased, which means the problem is "without the initial constraint manifold M". If the rule is selected and executed, after the "MODIFY" action, the transversality condition for To_term will be changed from "diff_M_over_To-diff_Theta_over_To-H" to "-diff_Theta_over_To-H".

IDSOC is implemented in VMS on a VAX 11/780 computer. It is composed of 98 rules. The distribution of these rules according to their functions is summarized in Table 2.

Table 2   Functions and distribution of rules

rule type	number	function
filter rule	16	create core rule LHS
core rule	34	find a solution
interface	48	build interface

5.4 Modifying knowledge base

IDSOC can also modify its knowledge base which contains the most information to deal with optimal control problems. Therefore, the particular solution may not exist in knowledge base when a linear control problem is to be solved. Usually, IDSOC can determine the solution structure and transversality conditions, and suggest the needed equation types. But for the linear control problem, we need the expert system to provide the information about solving the state feedback matrix K. In this case, the new production rules can be created to deal with this specific problem and to satisfy the need of the user. Being similar to IDSCA, in the OPS5, by using the action of "build", we can construct new specialized rules.

5.5 Imprecise knowledge representation

The rule-based expert system usually works in the domains where the conditions and conclusions are rarely certain. To guarantee the reliability of the conclusions, the certainty computing procedure is built in the expert system.

The expertise of IDSOC is in the domain of solving optimal control problem. To a large extent, the reliability of decisionmaking is associated with the certainty of input information provided by the user. For example, the value range of terminal time t_f and controllability of each state vector x_i ($i=1,2,......, n$) may affect the conclusions on Riccati Equation types.

Therefore, we use certainty factors to help represent the imprecise knowledge and indicate the reliability of decisionmaking. As an illustrated example, let us consider the handling of the uncertainty of input information about t_f and x_i. For this problem, three types of certainty factors are embedded in the certainty-computing procedure, that is, the certainty factor for t_f, the certainty factor for each state vector x_i ($i=1, 2, ..., n$) and the overall certainty factor. Each of the former two factors is composed of two parts: the input certainty factor and the corresponding rule's antecedents, while the overall certainty factor contributes to the manipulation of computation of the former two.

In the processing, when a user can not provide the precise input information, a menu-driven table will be shown to indicate the range of certainty factors corresponding to the related event, and ask the user to rank the certainty factor. Once all certainty factors associated with the related input information are given, the overall certainty factor can be calculated. The computing procedure is given below (Winston, 1984):

1. According to the rank of certainty factors given by the user, the values of C_{xi} and C_{tf} are assigned to the certainty factor of state vector x_i and terminal time t_f, respectively.

2. Corresponding certainty ratios are calculated by the following formula:

    $$R_{xi} = C_{xi} / (1 - C_{xi}) \qquad\qquad (5.3.1)$$

    $$R_{tf} = C_{tf} / (1 - C_{tf}) \qquad\qquad (5.3.2)$$

3. The overall certainty ratio R can then be obtained from R_{xi} and R_{tf}:

    $$R = R_0 \, (R_{xi} / R_0) \, (R_{tf} / R_0) \qquad\qquad (5.3.3)$$

    where R_0 is the prior certainty ratio that is from the expertise estimation about the relationship between the input information and the possible output results.

4. Finally, the overall certainty factor C can be calculated as:

    $$C = R / (1 + R) \qquad\qquad (5.3.4)$$

The next step is to choose the suitable output results based on the prior certainty factor C_0 and the obtained overall certainty factor C. The detailed description about the computation of certainty factors is given by Winston (1984).

5.6 User-friendly interface

The interface of IDSOC consists of two parts, that is, the input interface and the output interface. The input interface is in the "menu-driven" style. The user is asked to give the necessary information for the relevant problem at the various terminology chosen by the user. For example, the problem structure may be classified into two forms, that is, mathematics and performance measure.

The output interface returns the results of the decision to the user. The following contents are provided:

(a)   input information,
(b)   definition of functions used in this solver,
(c)   results of decisionmaking, and
(d)   procedure of rule-execution.

5.7 Integration environment

The Distributed Intelligence System architecture has been successfully applied to process control problems. For example, an intelligent optimal control system has been implemented, which consists of three expert systems: IDSOC, IDSCA and a Meta-system, a symbolic manipulation program, namely SMP as well as several numerical computing routines. As described in Figure 5.4, four numerical computation packages have been developed to handle system dynamics analysis, solving differential equations and matrix operations, and designing optimal control law as well as performing time-domain response simulation. SMP, a Lisp program, which is implemented for symbolic manipulation, can communicate with IDSOC and IDSCA as well as numerical programs through the meta-system. IDSOC and IDSCA are two symbolic reasoning systems. SMP (Symbolic Manipulation Program) is a symbolic manipulation program, which is developed to execute symbolic mathematical manipulation and processing based on the decision-making results provided by IDSOC.

Moreover, SMP can also be used to solve the purely mathematical problems separately.

This intelligent optimal control system is developed to handle various complicated optimal control problems, which are hardly solvable by hand or even by the CAD technique. Through the user-friendly interface, the system can receive the needed information from the user, perform heuristic search and quantitative computation, and then provide the design results quickly both on the screen and from a printer. The intelligent optimal control system shows the potential of both symbolic processing and heuristic reasoning for complex problems in optimal control. Besides the important features of IDSOC, it can coordinate numerical computation procedures with symbolic reasoning processes, and it integrates different expert systems in the intelligent optimal control system and utilizes knowledge architecture to control and manage knowledge.

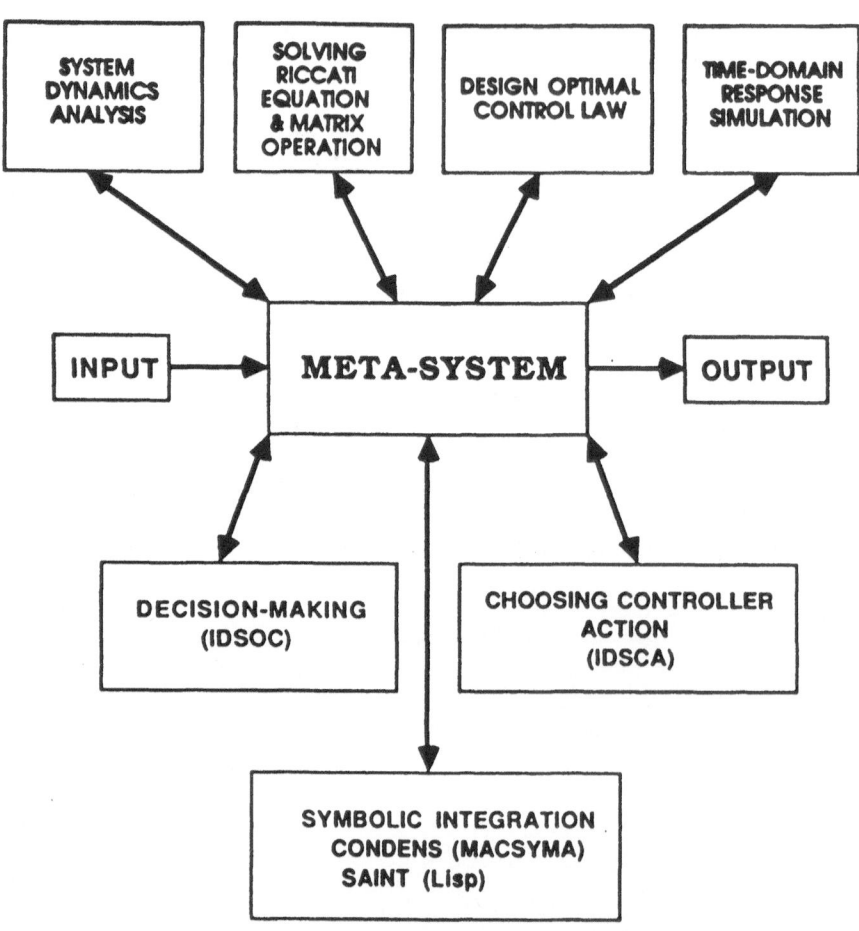

Figure 5.4 Intelligent optimal control

6 Pulp and Paper process control

6.1 Pulp and paper process control

Pulp and paper industries plays a very important role in Canadian economical development. As the largest manufacturing industry in Canada, most of these companies produce newsprint paper. Traditionally, we make the profits from producing the newsprint paper by means of high volume and low production cost. However, we are now facing the international marketing competition in low production cost newsprint paper, especially from the South and the Tropics. This has pushed our industries towards higher quality paper products, and greatly stimulates the research and development for pulp and paper process control. Pulp and paper processes are very complicated, in which the advanced technology and operation experience are highly concentrated. The fundamental requirements of pulp and paper mills are:

1.   good product quality,
2.   low production cost,
3.   high productivity,
4.   efficient manufacturing operation (such as short operation period),
5.   saving energy,
6.   operation safety and equipment protection,
7.   environment protection, and
8.   utilizing the most advanced technology.

Paper mills produce products which range from fine papers to corrugated cardboard. The product range requires that the basic raw material of paper has fibre characteristics to meet the specific grade production requirements. Chemical pulping digestion processes have been widely applied in paper mills. However, the effective control for such processes is still an unsolvable problem. An obvious phenomenon in papermaking processes is time delay. Time delay is also one of the important characteristics of non-minimum-phase systems. The processes with time delay are often very difficult to control. Meanwhile, papermaking process control always deals with multivariables. For example, in the control of a paper machine, basis weight and moisture are both coupled strongly. Other additional variables such as opacity or color control can also be involved. On a headbox, temperature, pulp concentration, pulp flow rate, and air pressure affect paper quality simultaneously. In summary, important characteristics of pulp and paper processes are:

1. batch process,
2. multivariable,
3. time-variant,
4. distributed parameter,
5. time delay,
6. nonlinear,
7. periodic operation procedure,
8. ill-formulated model, and
9. stochastic process.

The control problems involved in pulp and paper processes are usually ill-structured, and difficult to be formulated. In such processes, mathematical modelling is not amenable, and purely algorithmic methods are difficult to use. However, AI techniques provide programming methodology for solving these ill-formulated engineering problems.

In pulp and paper processes, operating conditions are frequently changed based on different production criteria. There exist many periodic operation procedures. Expert systems are suitable for use in such an environment. The stochastic occurrence of operational faults requires emergency handling in pulp and paper processes. Intelligent fault diagnosis systems are very powerful in dealing with such complex situations.

6.2 System configuration

As an academic research project to be developed for the real industrial plant operation, I would like to "plan big, start small", i.e., we first construct a basic framework for plant-wide intelligent control system for pulp and paper processes. This system is a "Front-to-End" frame, in which a digester in pulp process is selected as the "Front" subprocess, while a paper machine as the "End" subprocess. These two process units are not only the most important equipments in production, but also the most difficult ones to control. An intelligent scheduling system will be developed to ensure operation planning, quality control, production safety and environmental protection. An integrated intelligent system is implemented as the software integration platform for the above intelligent control systems (Figure 6.1).

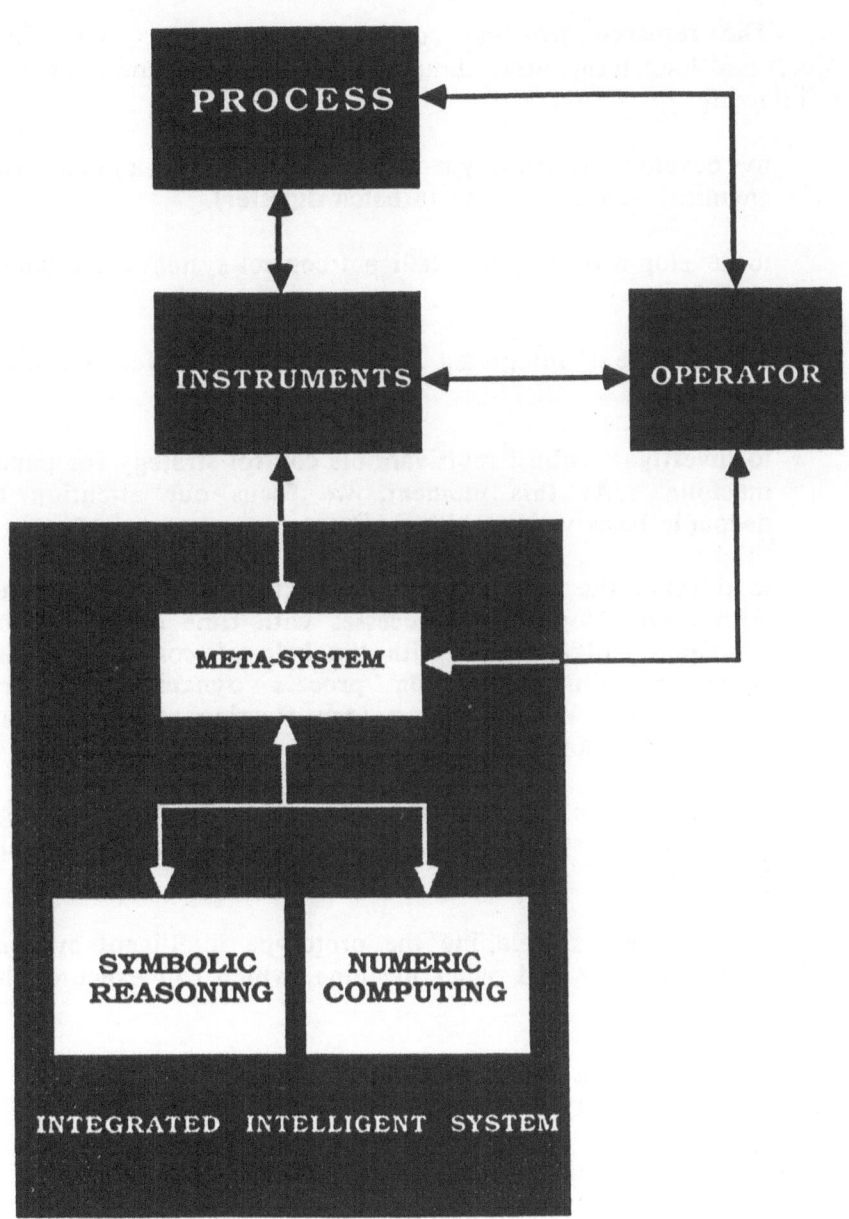

Figure 6.1 Intelligent process control

The research projects consist of two parts: immediate projects and long term ones. Immediate projects are summarized as the following:

1.   to develop a prototype intelligent control system for chemical pulping process (a batch digester).

2.   to develop a prototype intelligent control system for a paper machine.

3.   to implement integrated intelligent system for software integration.

4.   to investigate robust multivaraible control strategy for paper machine. At this moment, we focus our attention to decouple basis weight with moisture.

5.   to develop the effective control strategy for the processes with time delay. The processes with time delay will be simulated and analyzed with the help of computer. The effect of time delay on process dynamics will be investigated. Various system identification techniques will be developed and utilized.

The immediate projects will pave the way for our long-term goal that is to develop plant-wide intelligent control for pulp and paper processes.

Also, we are developing the prototype intelligent systems for the above digester and paper machine, which can executes the following functions:

1. production planning and scheduling,
2. operation and quality control,
3. fault diagnosis, and
4. production safety and equipment protection.

6.3 Intelligent system for batch digester

High quality pulp is often produced by a chemical process under a batch operation, which utilizes cooking liquor to dissolve and to remove the lignin from wood chips for the production of chemical pulp. The kraft pulping process is used to produce chemical pulp with a desired degree of delignification (desired kappa number). A batch digester is a large chemical reactor where wood chips and cooking acid are charged and mixed together for the production of sulfite pulping. The purpose of the digester is to separate the wood fibers from the lignin (glue) by controlling temperature, pressure and chemicals. The digester cooks the chips in acid under controlled temperature and pressure. The acid, which consists of mixture of SO_2 dissolved in water and combined SO_2 in the form of magnesium (Mg), dissolves the lignin as well as leaves the fibers in pulp form.

When a product is manufactured under a batch operation, the desired quality of the product is only reached at the end of the batch cycle. PID (Proportional-Integral-Derivative) or programming logic controllers cannot be used in the same way as in a continuous process. We have developed an integrated intelligent control system to provide the operators an estimation of the total time required for the cooking cycle to achieve a target kappa number for given pulping conditions. A symbolic reasoning system is being developed to reduce the process model mismatch by updating the model parameters, which allows to establish qualitative models and quantitative models, and helps to obtain a much better control. It can also be used to detect unexpected events and faults during process operation.

Nine inputs variables are identified as the major variables that affect the cooking process. They are: chip quality, chip load, maximum temperature, maximum pressure, cooking time, total % SO_2, free % SO_2, combined % SO_2, and pH. An expert system was developed to monitor these variables.

The second expert system was constructed based on operation procedures. Its knowledge base contains a series of instructions describing step by step for a whole single cycle.

The third expert system was developed to handle emergency operation, shutdown operation, as well as startup operation. It aims at providing the operator the decision support related to the above situations.

A numerical simulation program was implemented to predict the S-factor required for the process to achieve a specified cooking degree.

These three expert systems and the numerical computing program were under the control by a meta-system. The meta-system also interfaces the integrated intelligent system with the human operators.

6.4 Paper machine intelligent control

Paper machine plays a key role in papermaking process. A good control of paper machine is extremely important. There are various paper machines. Almost all paper making processes are continuous, but have to shutdown and startup periodically due to washing wire, changing some parts (such as wire, felts, calenders, etc.). In some cases, a paper machine needs to produce different kinds or different grades of paper.

An intelligent control system was developed for paper machine which provides such functions as process operation at an optimum economic level, minimized rejections for out-of-specification product quality, targets adjusted automatically as process conditions change, and decision support for operators when emergency occurred.

This intelligent control system mainly consists of two parts: model-based control system and knowledge-based control system (Figure 6.2). The model-based system contains numerous effective control strategies for the process with nonlinearity, time-variant, time delay, and coupling property, such as robust, optimal, predictive, bilinear, reliable, fault-tolerant and fuzzy control strategies. They can deal with the normal process optimization, simulation and control using the process mathematical model. Knowledge-based system can handle shutdown, startup, and emergency operation, faults diagnosis, control strategy selection, control target value adjusting based on operation knowledge and product quality requirements.

Besides the applications in pulp and paper process control, my research laboratory is developing several integrated intelligent control systems for bioreactors and HVAC (Heating, Ventilating and Air Conditioning) Process. Figure 6.3 shows the intelligent bioreactor control system structure, while Figure 6.4 demonstrates the application of integrated intelligent system to HVAC process control.

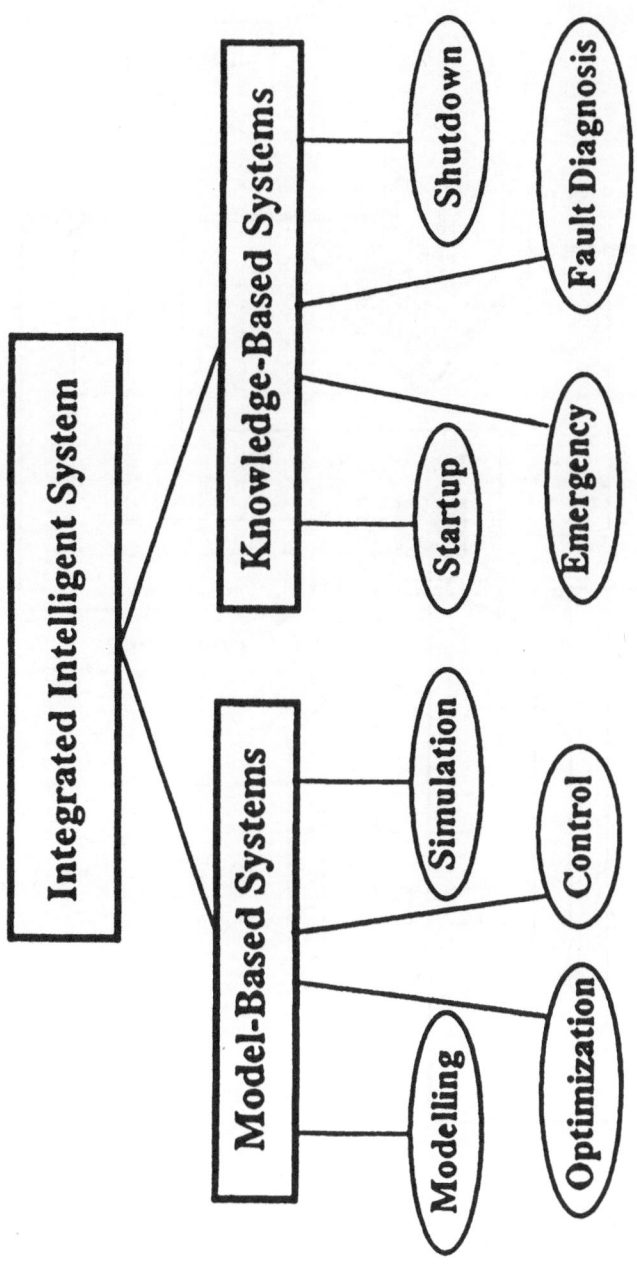

Figure 6.2   Intelligent control system for paper machine

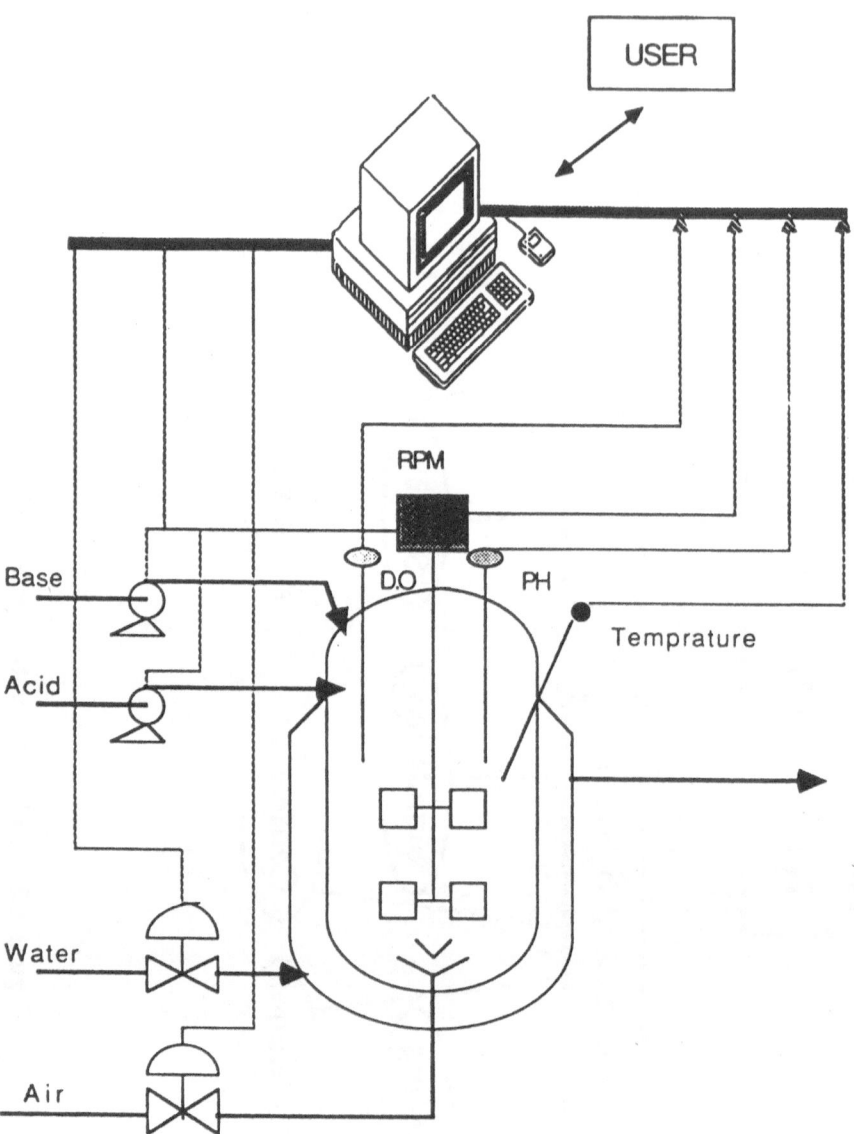

Figure 6.3 Intelligent control of bioreactor

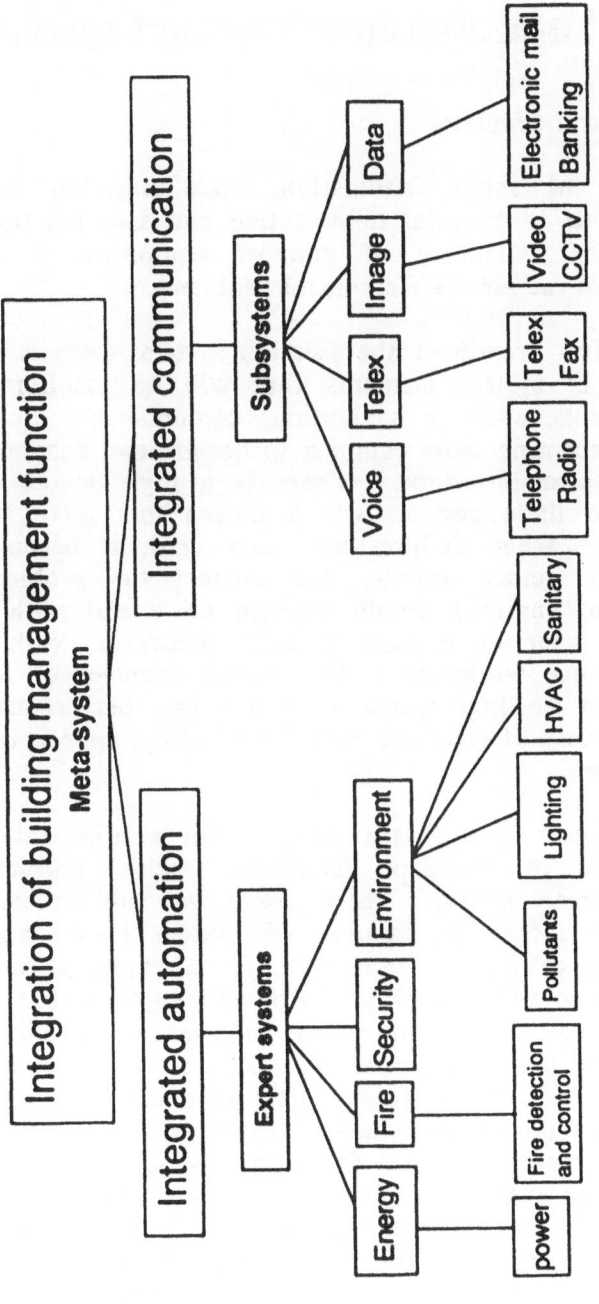

Figure 6.4 HVAC intelligent control function

7 Intelligent maintenance support system

7.1 Air-traffic control problems

Maintenance of radar, navigation, communication and environmental systems is essential to effective and safe aviation. The availability and reliability of support equipment is of paramount concern to the air-traffic control problems.

Due to the fast growth of the aviation system, there is an increasing number of system elements that will be monitored. These system elements, such as navigation, communication, and radar services are becoming more complex in design, and generally becoming simpler to maintenance. There is a high degree of reliability built into these systems which means that there are infrequent failures. When failures are infrequent, it becomes difficult to quickly detect, isolate, and correct the problem. Previously, a system engineer would become an expert with a particular type of radar or navigation aid. However, with a centralized maintenance workstation, the system engineer of the future will need to be knowledgeable about key performance parameters for a variety of facilities in order to direct appropriate maintenance activities.

The maintenance operations automation for aviation safety involves a number of complex technical, social, political, economical, and human issues. There are many sub-problems associated with the increasing number of system maintenance elements. At times of multiple outages, system engineers may experience cognitive overload and performance degradation may result. There could be a failure to monitor all available information or resolve conflicting constraints in a tactical situation. As previously mentioned, due to technological advances, equipment reliability has been typically increased. However, more reliable systems means that systems do not fail as often as in the past. These scarce problems make it difficult to quickly diagnose and determine appropriate maintenance actions due to unfamiliarity with particular problem types. With the aviation system, there is a high cost associated with slow response time and mistakes, particularly since human lives are involved. Due to the absence of an integrated workstation and the abundance of monitor screens in the current configuration of the system engineer's workstation, it becomes difficult for the system engineers to simultaneously manipulate all relevant information to obtain optimal solution quickly.

7.2 Expert systems for maintenance

Expert systems provide a programming methodology for solving ill-structured problems which are difficult to handle by purely algorithmic methods. An expert system is also a computer program that acquires the knowledge from human experts and applies it to make inferences for the user with less training or experience in solving various problems. The separation of the database, the knowledge base and the inference engine in the expert system allow us to organize the different methods and domain expertise efficiently because each of these components can be designed and modified separately. The experience gained from developing expert systems has shown that their power is most apparent when the problem at hand is sufficiently complex.

The development of expert system for air-traffic control facilities will capture the knowledge from retiring domain experts, and will enhance the productivity of system engineers by providing decision aiding tools with advanced capabilities. Since some of the expertise is vague, scarce, and dispersed, it should be preserved, made precise, focused, and continuously applied. The development of a knowledge base for intelligent maintenance support of air-traffic control equipments will enable the codification of existing system engineers' expertise before this important expertise leaves the aviation fields. Once captured, the knowledge can be efficiently applied on a continuous basis via an expert system to enhance the decision making productivity of both novice and experienced system engineers. This knowledge can also distributed to the other similar applications. The primary applications of expert systems to system engineers' environment can be summarized as follows:

1. Diagnosing, interpreting as well as monitoring problems.
2. Choosing analysis and modeling tools.
3. Selecting facilities configurations.
4. Planning for predictive maintenance and refurbishment.
5. Capturing, duplicating, as well as transferring expertise.
6. Training novices.

Figure 7.1 presents an example of expert systems used in air-traffic control.

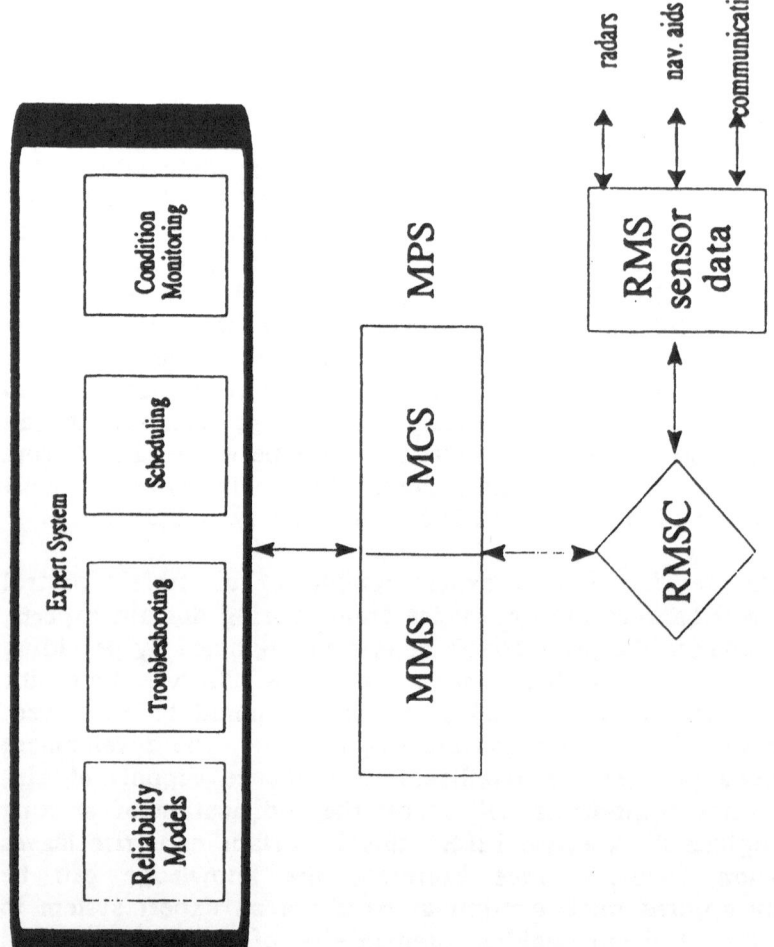

Figure 7.1 Expert system in air-traffic control

7.3 Interdisciplinary research methodology

The primary objective of this research is to use an interdisciplinary approach to identify opportunity for the application of existing and emerging technologies to facilitate automation of maintenance support operations for air-traffic control facilities. The main objective is to use the framework of a meta-system to integrate various expert systems, numerical computing modules, resource allocation algorithms, and failure prediction models into an intelligent decision aid for system engineers. The meta-system would be a supervisory expert system that will contain the "knowledge about the knowledge". In a tactical decision making situation, the Intelligent Maintenance Support System would activate the appropriate symbolic reasoning system, algorithm, numerical processing package, or mathematical model and guide the system engineers in the problem resolution. Intelligent Maintenance Support System (IMSS) integrates task specific knowledge, and assists the system engineers by serving in an advisory role. Such an intelligent assistant will reduce the stress associated with the system engineers cognitive workload.

Our project consists of several phases. Beginning at May 1989, we have successfully accomplished the objectives in organizing an interdisciplinary research team, defining problems, identifying system functions and tasks, doing literature search, visiting several fields to interview system engineers, evaluating and selecting expert system development tools, constructing prototype Intelligent Maintenance Support System architecture, developing a paradigm for predictive maintenance, and so on. For example, in order to facilitate accomplishment of the primary objective, a set of well defined tasks were performed. The task descriptions are as follows:

Task 1   Identification of the functions and systems, subsystems, and facilities to be supported by IMSS.

Task 2   Identification of the decision making tasks. Tasks which were candidates for performance aiding were listed and candidate aiding approaches were selected (expert systems, database searches, algorithms, cueing/prompting, or task step guidance). For tasks selected as candidates for expert system aiding, the sources for knowledge base generation were identified.

Task 3   Based on the output from tasks 1 and 2, potential application areas for the meaningful use of artificial intelligence within maintenance support system were identified.

Task 4   Based on expert system tools evaluation and through technical interviews, a framework for the Intelligent Maintenance Support System (IMSS) was developed. The technical requirements, such as expert system environment selection, meta-system function analysis, system interface requirements, etc., were defined.

Task 5   Supportability for the IMSS on existing hardware and software platforms was determined. Candidate expert system building tools were evaluated to determine their suitability for prototype of the meta-system.

Our interdisciplinary research team organization includes the following six members:

Dr. C. Theisen: Human Factors
Dr. J. Luxhoj: Industrial Engineering
Dr. M. Rao: Intelligence Engineering, Control
Dr. Adi Ben-Israel: Operation Research
Mr. S. Burgess: Computer Science
Mr. D. Cuthbert: Psychology

7.4 Development of application systems

Maintenance of critical systems and components is essential not only to the operational success of the air-traffic control, but to any industrial, service, or government organizations. Over the years, there have been a number of maintenance paradigms developed that model the behavior of items subject to mechanical or electrical failure. Luxhoj [1988] presents a survey of the literature on maintenance modeling and proposes an organization structure to assist in understanding the multi-faceted features of maintenance paradigms. This structure involves the development of seven general categories which are: system states, decision actions, planning horizons, system knowledge, system objectives, solution methodologies, and modeling features.

After analyzing the qualities of the problem domain, it appears that some advanced decision support concepts could be implemented with current technology. The central research idea espoused by the team is the development of a meta-system (a supervisory intelligent system that would activate individual expert systems, algorithms, or numerical computation modules, which contain localized knowledge). The Intelligent Maintenance Support System (IMSS) is an integrated intelligent system [Rao et al., 1987; 1989] that integrates task specific knowledge, and assists system engineers by serving in an advisory role. Such an intelligent assistant will reduce the stress associated with the system engineers cognitive workload.

We had limited access to system engineers during our early work of this research. Essentially, system engineers' major responsibilities are as follows:

1. Prevent/reduce service outages.
2. Monitor the state of the air traffic control facilities.
3. Respond to system problems.
4. Manage the technician workforce.
5. Prepare administrative reports.

As a result of interviews with system engineers at several different sites, the following activity trees were constructed:

1. Equipment hard alarm.
2. Equipment soft alarm (i.e. failure of noncritical component, alarm thresholds, intermittent alarms).
3. Remote preventive maintenance.
4. Remote certification.
5. Coordinate site preventive maintenance.
6. Coordinate inspections.
7. Aircraft accident/incident and emergency coordination.
8. Modification coordination.
9. Ad Hoc activity.
10. Shift changes.
11. Failure of utilities.

These activities trees analyses are very valuable in structuring the knowledge base for the prototype of the IMSS, since the decision trees formalize the heuristic maintenance procedures. Potential IMSS application areas are depicted in Figure 7.2.

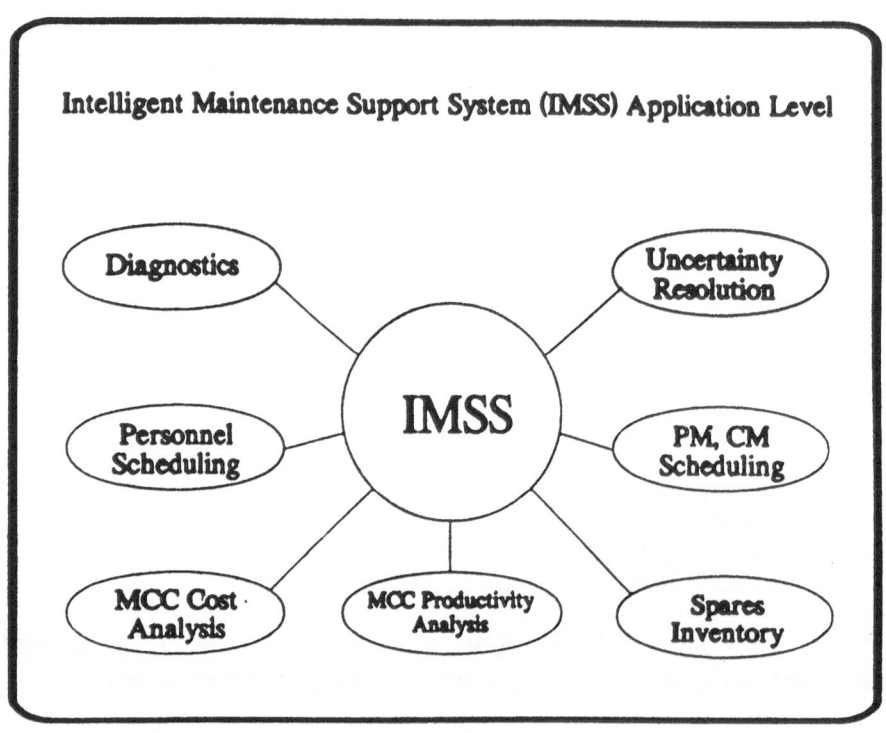

Figure 7.2 Potential IMSS application areas

Integrated Intelligent System [Rao et al., 1987; 1989] is applied to the IMSS, which is a large knowledge integration environment, and consists of several symbolic reasoning systems and numerical computation packages. These software programs may be written in different languages, and may be used independently. They are under the control of an intelligent supervisory system, namely meta-system. Meta-system manages the selection, operation and communication of these programs. The key issue in implementing IMSS is to organize a meta-system, which executes the six main functions, such as to coordinate symbolic reasoning and numerical computation; to distribute knowledge; to acquire and integrate new knowledge; to handle conflicting events; to provide the possibility for parallel processing in IMSS, and to standardize communication information. For the more details about the meta-system, please refer to the references [Rao, et al., 1987; 1989]. Figure 7.3 shows the application of the integrated intelligent system to real-time maintenance support for the air-traffic control facilities.

7.5 Summary

A new approach to use distributed intelligent system techniques for air-traffic control facilities maintenance is developed, in which the Intelligent Maintenance Support System (IMSS) architecture is proposed. IMSS consists of several expert systems and numerical computing packages as well as a meta-system. The IMSS can serve in an advisory or consultant capacity, making the best use of both human and machine intelligence.

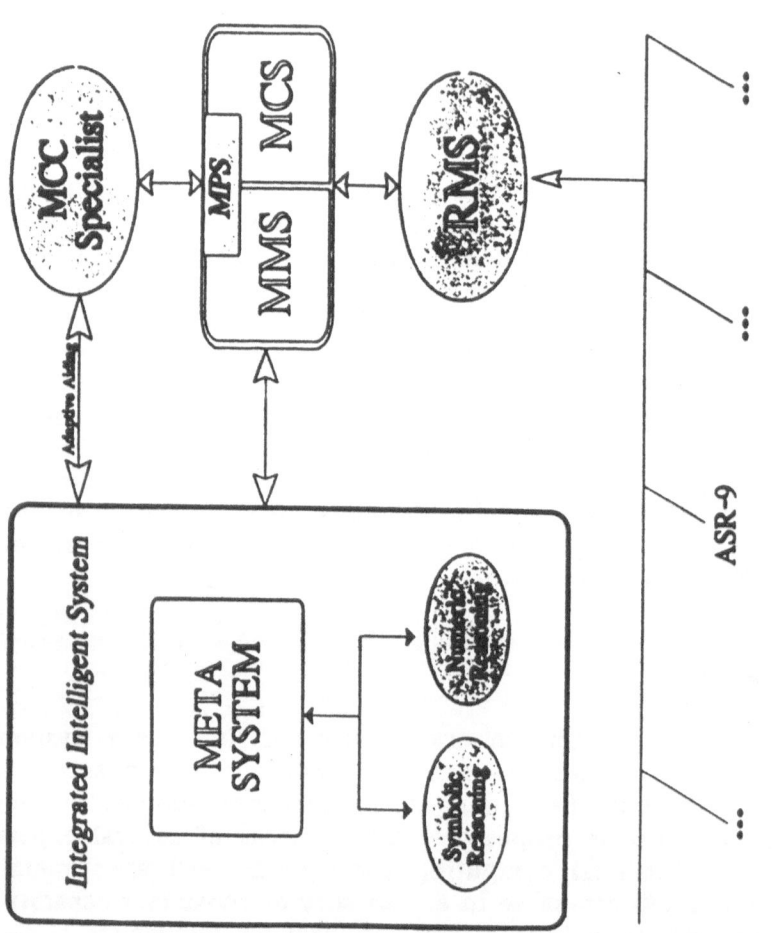

Figure 7.3 Integrated intelligent system for maintenance facilities

8 Gear integrated manufacturing system

8.1 Integrated manufacturing

The growing complexity of manufacturing processes and the need for higher efficiency, greater flexibility, better product quality, and lower cost have changed the nature of industrial practice. Meanwhile, the application of computers has allowed the implementation of more advanced modeling techniques [Luxhoj and Jones, 1988].

Computer Integrated Manufacturing (CIM) is generally acknowledged to be the next phase to CAD/CAM development. It holds the key to an order of magnitude savings in production time, labor and tool cost, reduction in human errors, and a more versatile and quality product. With CIM, product design, manufacturing, testing and production management are integrated into a unified system which is a very comprehensive and sophisticated environment in terms of software and hardware. Numerous efforts have been made to investigate the architecture of an integrated software system for CIM [Su and Lam, 1990; Bourne, 1986; Roboam, Sycara and Fox, 1990], since it is recognized as the key issue for the integration of so many complicated subsystems. Intelligent manufacturing systems research are exploring the potential driving force for our next generation manufacturing facility [Bourne, 1986; Lu, 1989].

In this section, a new framework for an intelligent manufacturing system is proposed to solve the problem encountered above.

In recent years, there have been numerous attempts to add "intelligence" to the flexible manufacturing environment. There has been a proliferation of "expert systems" developed for many of the operational areas associated with manufacturing, such as scheduling, materials handling, quality control, maintenance, and productivity. Expert systems provide a programming methodology for solving ill-structured problems which are difficult to handle by purely algorithmic methods. An expert system is also a computer program that acquires the knowledge of human experts and applies it to make inferences for the user with less training or experience in solving various problems. The segregation of the database, the knowledge base and the inference engine in the expert system allows us to organize the different models and domain expertise efficiently because each of these components can be designed and modified separately.

8.2 System organization of GIMS

The integrated intelligent system has been successfully applied to several complicated applications. Here, we only present a gear integrated manufacturing system briefly as an example.

The architecture for integrated manufacturing systems based on the theory of integrated intelligent system has been developed [Rao, Cha and Zhou, 1991], as shown in Figure 8.1. Its feature is to establish a meta-system for intelligent integration of existing CAD, CAM and CAT subsystems which are independently developed and used in a different language environment.

A Gear Integrated Manufacturing System (GIMS) is developed to carry out gear design, manufacturing and testing as well as fault diagnosing within one integrated environment. The meta-system plays a key role in GIMS for integration, management, coordination, control and communication. Written in TURBO-PROLOG, it can take commands and tasks from users through an interface, and allocate the tasks to different subsystems and monitor the process carried out in subsystems, and can also manage the whole system in many aspects. The following is a detailed discussion about the structure and functions of the integrated software system using GIMS as an example.

The ICAD (Intelligent CAD) subsystem engages product design. In GIMS, the ICAD subsystem consists of an expert system for gear design which is written in an expert system tool INSIGHT+2, and several programs written in FORTRAN and QUICK-BASIC for gear analysis and calculation. ICAD also can make mechanical drawing.

The ICAM (Intelligent CAM) subsystem is in charge of process planning, and programming for a Numerical Control (NC) machine. In GIMS, ICAM is developed using the M.1 expert system tool for gear process planning and programming, and in FORTRAN for some analysis and calculation. ICAM can provide process programming for an NC machine based on the information supplied by the ICAD subsystem.

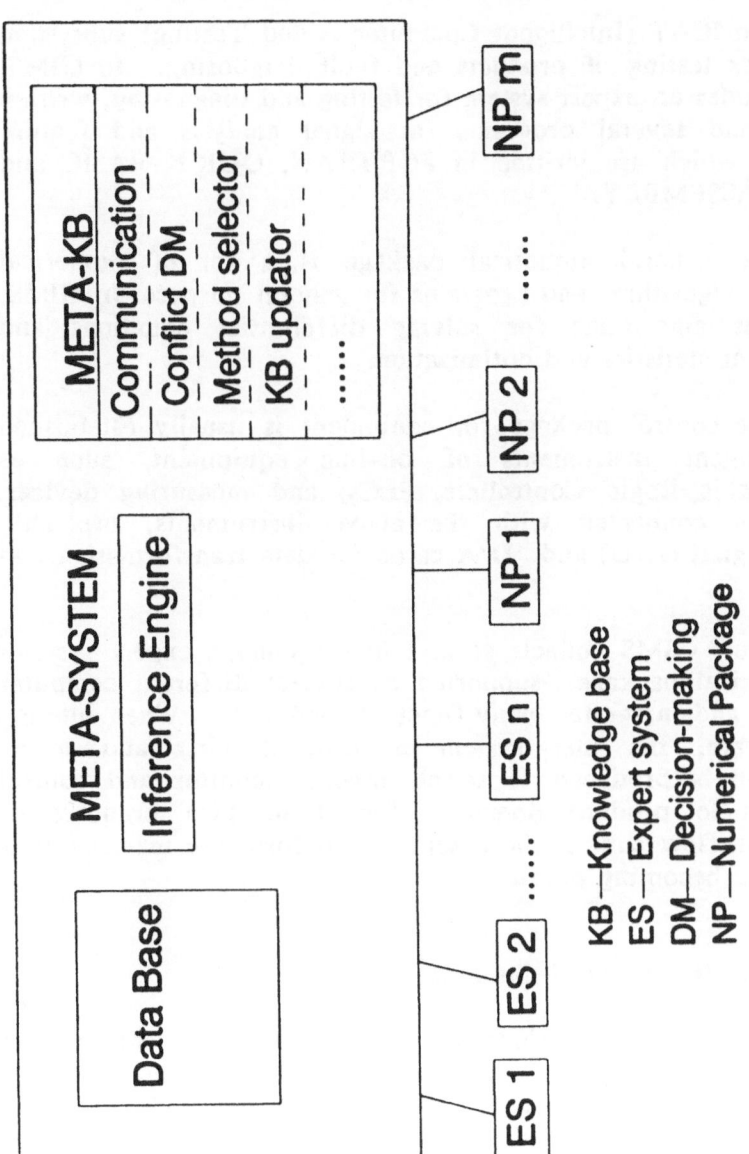

Figure 8.1 Architecture of integrated intelligent manufacturing

The ICAT (Intelligent Computer-Aided Testing) subsystem is for error testing of products and fault diagnosing. In GIMS, ICAT includes an expert system for testing and diagnosing, written in M.1, and several programs for signal analysis and feature extraction which are written in FORTRAN, QUICK-BASIC and MACRO ASSEMBLY.

The general numerical package is a set of numerical calculation algorithms and programs for general purpose. In GIMS, it contains algorithms for solving differential equations and packages for statistics and optimization.

The control package for equipment is usually established for intelligent instruments of on-line equipment, such as Programmable Logic Controllers (PLCs) and measuring devices. GIMS, not connected with the above instruments, has only Analog/Digital (A/D) and D/A cards for data transformation and exchange.

Thus, GIMS collects several heterogeneous expert systems and numerical packages, supported by several different computer languages, and integrates many functions and tasks. When running the software, the meta-system is invoked first and meta-information is produced to select, invoke, monitor and control subsystems to produce domain information. Two branches of information flow control the equipment to form the material flow with blanks becoming products.

8.3 Modularity of knowledge bases

By different functions, knowledge can be categorized into "domain knowledge" and "meta-level knowledge". The former is usually defined as facts, law, formula, heuristics and rules in a particular domain of knowledge about specific problems, whereas the latter is defined as knowledge about domain knowledge and can be used to manage, control and utilize domain knowledge. There are some differences between the two categories of knowledge. For example, meta-knowledge possesses diversity, and covers a broader area in content and varies considerably in nature. Also, it has a fuzzy property of more uncertainty than fact. Hence, the representation and knowledge-base structure of meta-knowledge should have their own characteristics.

8.4 Communication in GIMS

In GIMS, the meta-knowledge is represented with a combination of production rules and frames. The meta-knowledge base is constructed by modules. Serving as a manager, the meta-knowledge-base-management-module is the only one connected to the inference engine and passes its commands for reasoning to other proper modules. The subsystem-management-module can possess the knowledge about subsystems, such as the language environment, the knowledge representation techniques, the inference mechanisms, the functions and control strategies. It is only through this module that the meta-system is able to communicate with different subsystems and to choose the right one to perform a certain task. In GIMS, the database-management-module is responsible for communication between the meta-system, written in TURBO-PROLOG, and the database, DBASE III. The knowledge about the data structure and functions of DBASE III and PROLOG is stored in the module for data transformation and data file processing. The modules in the meta-knowledge-base are independent; hence, it is easy to modify, delete and add any module without disturbing the others. All modules connect to each other through the management-module, and can share knowledge with each other. This makes the meta-knowledge-base more flexible and efficient.

As a large integrated knowledge environment, GIMS will face two barricades for implementation with a single computer. First, due to the limitation of the on-board memory space, the whole system may not be able to load in all together at the same time. Second, due to the heterogeneity of the subsystems developed in different languages, the system may not be linked together as a whole and loaded in at the same time. Therefore, a technique, namely covered-structure, can be adopted to solve the problems. The basic strategy of the technique is to invoke the subsystems in turn by an order, controlled by the meta-system, in accordance with the memory space and the nature of the subsystems, then find the way to allow information exchange between the invoked subsystem and the ones at rest.

GIMS adopts the covered-structure technique and solves the communication problem in the following manner. There are two ways for communication, namely, direct communication and indirect communication. The direct communication method is employed for those programs which are used frequently and require quick access to the meta-system in order to transform data in time. Under the conditions of enough memory and compatibility of the languages, the programs may be compiled and linked with the meta-system and reside in memory for the whole process. In this way, the meta-system can communicate with the program directly. For example, a real-time control system for machine tools needs to get information about states of the machine within a second, from sampling signals of vibration or noise on the machine to obtaining the results calculated by a statistics program. The communications among the meta-system, numerical package, measuring devices and the control system have to be very fast. In such situation, the statistics program and the sampling software should communicate with the meta-system directly to guarantee the control system obtaining necessary data in time. Through standardization, the meta-system can accept the data from the programs linked with it, and fast communication can be achieved.

The indirect communication is realized through a "transformation station", namely a static blackboard (SB). The meta-system can invoke some executable programs by using the predicate SYSTEM(" ") in PROLOG. The results from the running programs are then sent to the SB, and processed there into standard forms for calling by other subsystems through the meta-system. For example, the diagnosis expert system of ICAT in GIMS communicates with other parts of the system in such manner. The procedure of indirect communication takes several steps as the following:

1. Based on the meta-knowledge, the inference engine sends program P1 the instruction that program P2 needs data from it and passes the language type of P2 to it.

2. The inference engine invokes P1, P1 starts to run.

3. The results from running P1 are written in the SB with a file name, such as 1.DAT, and informs the meta-system.

4. The meta-system identifies the data file by the name 1.DAT and standardizes it, then saves the results in S1.DAT.

5. The meta-system sends S1.DAT to program P2.

Indirect communication is complicated and slow. In GIMS, data to be written into the SB should be prepared in a certain structure. An example is presented below:

Transformation Symbol: "FM"
Variable Name: "Conclusion"
Logical Relation: "Equal to"
Value: "1024.6"
Certainty Factor: "cf90"
Data Feature: " . "

The data structure includes six segments. The Transformation Symbol indicates the two languages of the programs between which the data is going to be transformed. For instance, "FM" means the data will be transformed from a program written in FORTRAN to a program written in M.1. The Logical Relation stands for the relationship between the variable name and its value. For example, it may be "greater than", "less than" , "equal to" or "belong to" etc. The Certainty Factor (CF) stands for the degree of facts being true. In the above example, the certainty of the value "Conclusion = 1024.6" is "90%". The term of Data Feature refers to a special data structure that some languages may require (e.g. M.1 requires a dot . at the end for all data).

The procedure of standardizing the data in the SB takes two steps. First, the meta-system identifies the two languages from the transformation symbol. Then, based on the meta-knowledge about languages, processes the data into standard form acceptable to the receiver.

A commercial database, DBASE III, is adopted in GIMS, thus the implementation of communication between TURBO-PROLOG and ,DBASE III is an important content in the database-manage-module of the meta-knowledge base. The following is a simple description on transformation between TURBO-PROLOG and DBASE III.

Since TURBO-PROLOG can only recognize facts in symbolic sentences, and whereas the database can only store records, the key issue for communication between them is how to convert the records in the database into symbolic form acceptable to TURBO-PROLOG. The predicate is designed for this purpose:

Database_To_Prolog(Fact_name,
                    Database_files_name, Begin, Amount)

Here, Fact_name stands for the fact name having been transformed to TURBO-PROLOG, Data_files_name denotes the file name in the database to be transformed, "Begin" stands for the first record number to be transformed to TURBO-PROLOG, "Amount" denotes the number of records transformed to TURBO-PROLOG. The transforming process using TURBO-PROLOG is as follows:

1. Identifying the file name in the database in which the data are to be transformed, opening temporary files for transforming process, reading in the starting address and record length.

2. Reading in attribute values of the records, establishing an attribute table for them by using the retracing and list processing function provided by TURBO-PROLOG.

3. Determining the position of first transforming record according to initial recording mark "begin".

4. Reading in all records by the order given in the attribute table, and transforming the records to facts by a series of predicate operations. The facts are saved in the temporary files.

5. Reading in facts from the temporary files to TURBO-PROLOG system. Then the facts are further saved into the dynamic database of the meta-system.

The above discussion only covers the communication among the meta-system, database, expert systems and numerical programs. The communication between the meta-system and equipment, such as the machining center, needs further investigation in the future.

8.5 Optimal methods selection

When there are several methods in the system to accomplish one kind of tasks, it is important to select the most appropriate one for a specific task in order to raise efficiency and robustness. For instance, in GIMS, there are four methods for gear fault diagnosis and three methods for gear shaving, and each of them has its own characteristics. Selecting the best one for a particular task will involve symbolic manipulating and reasoning, so the meta-system should adopt the expert system technique to solve the problem. In the meta-knowledge base of GIMS, there is a module to provide the knowledge about those methods which can be employed to deal with same kind of tasks.

Finally, we discuss the way to choose the optimum method for gear fault diagnosis. Four methods that apply to gear fault diagnosis are listed below:

1. Spectrum analysis method, written in FORTRAN and MACRO ASSEMBLY for numerical calculation (nc),

2. Time series method, written in QUICK-BASIC (nc),

3. Major component method, written in QUICK-BASIC (nc),

4. Parameter statistics method, written in FORTRAN and MACRO ASSEMBLY (nc).

The symbolic processing for the four methods is developed with the expert system tool M.1. To select the optimal method, an optimization model should be established to evaluate the methods. The following four factors are considered for the evaluation function SF(X).

1. Diagnosis precision of the methods, denoted by Dp

2. Diagnosis speed of the methods, denoted by Ds

3. Expressive effects of the method, represented by Ex

4. Convenience of the method in operating, appraised by Co.

Thus, the evaluation function is established as following:

$$SF(Dp,Ds,Ex,Co)= k1*Dp + k2*Ds + k3*Ex +k4*Co+ k5 \qquad (8.1)$$

Here,
Dp,Ds,Ex,Co ----- Appraising factor
k1,k2,k3,k4 ----- Weighting coefficients
k5   ----- Revising coefficient.

Some constraints may be posed in the model for a specific diagnosis task. For example, to diagnose the original fault of gears, the precision should be a certain value and the speed is less important. For on-line diagnosis, high speed is required and precision should be given a lower bound. The best method should have the highest value of SF(X) that also satisfies the constraints.

In Equation (8.1), the parameters k_i ($i=1,...,5$), Dp,Ds,Ex,Co need to be determined by experience and knowledge of human expertise. In the system, this heuristic is represented in the meta-knowledge base, and based on it, symbolic reasoning is carried out to determine the optimal solution.

The principal features of GIMS can be summarized as:

(1) The hierarchical structure of an integrated intelligent system is suitable to integrated manufacturing systems for its flexibility and easy management. The meta-system is adopted as the kernel of the system to integrate, manage, control and utilize subsystems. Although the meta-system is similar to ordinary expert systems, it has unique structure and functions. The usage of the modular structure of the meta-knowledge base is convenient for managing the meta-knowledge and efficient for inferring.

(2) System communication is carried out by direct and indirect means. The indirect communication adopts a static blackboard as transformation station which is an effective communication method suitable for hierarchical systems.

(3) Optimal selection of methods should be done with a combination of expert systems techniques and numerical optimization methods. A proper optimization model with an evaluation function should be established based on heuristics.

The configuration of the integrated intelligent system has also been applied to mechanical systems design. For example, an intelligent system for combustion chamber design is demonstrated in Figure 8.2.

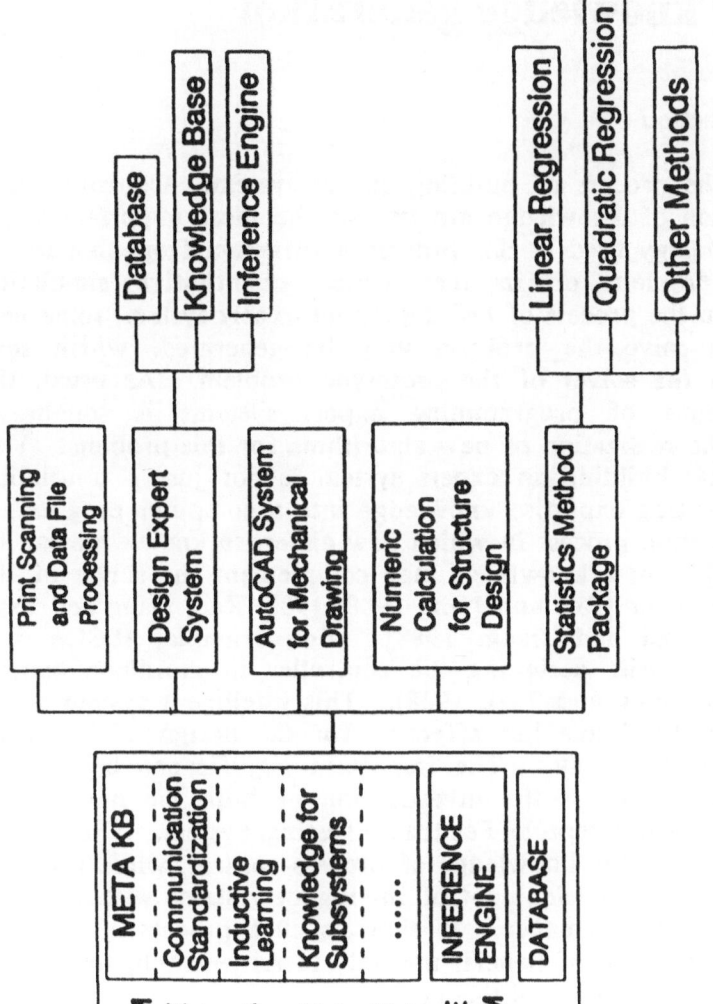

Figure 8.2 Intelligent system for combustion chamber design

9 New knowledge generation

9.1 Introduction

In the process of building expert systems, acquisition and representation of knowledge are two of the most important steps. The methodology used in this process is different from that in the prototype problem or in the related quantitative simulation program. In the process of developing an expert system, some new methods to solve the problem may be generated, which may complement the solver of the prototype problem. As usual, the new technique of programming expert systems is sought to guarantee the realization of new algorithms for this problem. This indicates that building an expert system is not just a translation from the existing expertise knowledge into a computer program, it is the production process in which new expertise knowledge can be acquired. The new knowledge may complement the solver of the prototype problem [Rao and Jiang 1988, 1989; Rao, Jiang and Tsai, 1988; Rao, Tsai and Jiang, 1988]. For example, IDSCA was developed to assist designing the controller in multiloop control system [Rao, Jiang and Tsai, 1988]. This intelligent system seems to be relatively simple but effective for the design of industrial process control systems. The important significance behind the practical application is the investigation of both the new design criterion and the Adaptive Feedback Testing System. The former may complement the knowledge of process control, while the later will stimulate the development of intelligent systems with the high reliability and performance. In short, such a process of building intelligent systems is of benefit not only to AI research, but also to the development of control theory.

In the following sections, I will introduce several new technological investigations that complement the theory and application techniques of both control engineering and artificial intelligence. Section 9.2 (a criterion to test non-minimum-phase systems) and Section 9.3 (a design criterion to choose controllers' action direction in multiloop systems) describe two new criteria for designing control systems, while other sections cover the contributions to artificial intelligence techniques.

9.2 Criterion to test nonminimum phase systems

A plant with a dead time or with a right half-plane zero is usually referred to as non-minimum-phase (Clark, 1984). However, those discrete-time systems whose zeros lie outside the unit-circle stability region are also called non-minimum-phase. We focus our attention on the latter.

Recently, much progress has been made on the control of non-mimimum-phase systems. Astrom and Wittenmark (1974) proposed some applicable algorithms. Goodwin and Sin (1981) made a new contribution on the convergence properties of the control algorithm. Clark (1984) proposed the idea of the self-tuning controller especially available for non-minimum-phase systems. Astrom et al. (1984) developed the research on the transmission zeros and explored the relation between unstable zeros and sampling period.

Although many criteria for the testing of minimum-phase systems have been developed (Jury, 1961 and 1964; Thoma, 1962, 1963; Xi and Schmid, 1985), those for non-minimum-phase systems are so few that they are much needed in industrial practice. Even though Jury (1964) had derived some rules to test the systems with unstable zeros or poles, they seem to be somewhat inconvenient for use in computer-aided control system design (CACSD).

A new and simple but effective criterion for determining a non-minimum-phase system by testing the coefficient of its characteristic polynomial is developed in the process of developing intelligent control systems. Since the method dose not require any calculation or tables, it is very useful in the CAD environment.

New Criterion

Consider an nth-order linear discrete-time system described by the following equation:

$$y(K)+a_1 y(k-1)+......+a_n y(k-n) \quad b_1 u(k-1)+b_2 u(k-2)+......+b_n u(k-n)$$

$$+q[e(k)+c_1 e(k-1)+....+c_n e(k-n)] \qquad (9.2.1)$$

Here k=...,-2, -1, 0, 1, 2,... . y is the output variable, u the control variable, q is the disturbance magnitude and (e(k) a sequence of independent normally distributed random variables between 0 and 1. Since we focus our attention on the stability of the system, $e(k)$ can be considered as zero for convenience of discussion. Therefore, we can rewrite (9.2.1) as

$$Y(z)A(z)=U(z)B(z) \tag{9.2.2}$$

in which,

$$A(z)=z^n+a_1z^{n-1}+...+a_{n-1}z+a_n \tag{9.2.3}$$

and

$$B(z)=b_1z^{n-1}+b_2z^{n-2}+......+b_{n-1}z+b_n \tag{9.2.4}$$

Criterion

If coefficients a_i ($i=1, 2, ..., n$) of (9.2.3) satisfy any one of the following conditions,

$$|a_1|>c_1=N!/[(n-i)!i!] \tag{9.2.5}$$

then at least one root of $A(z)=0$ lies outside unit circle.

Proof

For a polynomial equation defined by

$$P(x)=d_0x^n+d_1x^{n-1}+...+d_{n-1}x+d_n \tag{9.2.6}$$

in which d_0 is not equal to 0, the following fundamental algebra theorem exists between roots and coefficients if Equation (9.2.6) has solutions (Korn and Korn, 1961):

$$d_i/d_0=(-1)^i\Sigma[\text{product of roots (x) taken i roots at a time} \tag{9.2.7}$$

where the summation has $_nC_i$ terms.

Applying (9.2.7) to (9.2.3) gives:

$$a_1=(-1)^i\Sigma[\text{product of roots (z) taken i roots at a time} \tag{9.2.8}$$

where the summation has $_nC_i$ terms.

If all roots of A(z) lie inside unit circle, we have

$$|z_i| < 1 \qquad\qquad (9.2.9)$$

This implies that all products of the roots (z) in the summation of (9.2. 8) are of magnitude less than 1. Since there are $_nC_i$ terms in the summation of (9.2.8), we have for any i (i=1, 2, ..., n),

$$|a_i| <_nC_i \qquad\qquad (9.2.10)$$

In other words, if $|a_i| >_nC_i$ for any i (i=1, 2, ..., n), some roots of A(z)=0 must lie outside the unit circle of the z-plane.

A similar criterion can be obtained for B(z). It states that if any one of the following conditions is satisfied:

$$|b_{j+1}| >_{(n-1)}C_j |b_i| = [(n-1)!]^* |b_1| /[(n-j-1)!j!] \quad (9.2.11)$$

where j=1, 2, ..., n-1, then at least one root of B(z)=0 lies the unit circle, or the system is non-minimum-phase since unstable zeros exist.

<u>Illustration</u>

Example 1

Consider a second-order linear discrete-time system (n=2) defined by

$$Y(z)^*(z^2+a_1z+a_2)=U(z)^*(b_1z+b_2) \qquad (9.2.12)$$

Let $A(z)=z^2+a_1z+a_2=0$ $\qquad\qquad (9.2.13)$

We have the following two roots:

$$z=-(a_1/2)\pm[(a_1/2)^2-a_2]^{1/2} \qquad\qquad (9.2.14)$$

If $|a_1| = |z_1+z_2| > 2$, or $|a_2| = |z_1|^* |z_2| > 1$, at least one root of (9.2.13) lies outside the unit circle. This is in agreement with (9.2.5). Similarly, since

$$B(z)=b_1z+b_2=0 \qquad\qquad (9.2.15)$$

We know

$$z = -b_2/b_1 \qquad\qquad (9.2.16)$$

According to (9.2.11), we conclude that when $|b_2| > |b_1|$, $|z| > 1$ or the root of $B(z) = 0$ is outside the unit circle.

Example 2

The system below has an unstable zero at z -2.0,

$$Y(z)(z^2 - 0.7z) = U(z)(z+2) \qquad\qquad (9.2.17)$$

where n=2, $b_1 = 1$ and $b_2 = 2$.

Since $|b_2| > |b_1|$, this is a non-minimum-phase system as indicated by (9.2.11).

Example 3

The following system has two unstable zeros at 1.1 and 2.0:

$$Y(z)^*(z^3 - 0.2z^2) = U(z)^*(z^2 - 3.1z + 2.2) \qquad\qquad (9.2.18)$$

where n=3, $b_1 = 1$, $b_2 = -3.1$ and $b_3 = 2.2$.

Since $|b_2| > (n-1)^* |b_1|$ (j=1 in (9.2.11) and $|b_3| > |b_1|$ (j=2 in (9.2.11)), either criterion can be used to show that the system is non-minimum-phase.

Summary

A simple but effective criterion for testing non-minimum-phase systems is introduced. The criterion is a sufficient condition for non-minimum-phase systems. It is consistent with, but simpler than, Jury's method (1964). Because of its simplicity we suggest this criterion should always be tested first in determining non-minimum-phase for a system before applying more complicated criteria. This criterion is very useful in the procedure for designing computer-aided and knowledge-based control systems, particularly when it is applied to higher-order systems.

9.3 Criterion to select controllers' direction action

In designing a control system, a key decision is to design a controller in the following two phases:

1.   choose the suitable control law,
2.   choose correct direction of controller's action.

Usually, the work in phase 1 concentrates on the control theory, but that in phase 2 deals with both control theory and instrumentation. So far, among many reports on the multivariable control [Rosenbrock, 1969; MacFarlane, 1972; Owens, 1981; Pang and MacFarlane, 1987], the successful techniques for choosing the direction of the controller's action in a multiloop control system have not yet been developed, even though they are very important for industrial environment. In practice, industrial control engineers are normally concerned with the installation of instrumentation. The designers of process control systems always have to deal with the procedure of choosing the direction of controller's action. This directly affects the process operation and safety. The procedures often vary, depending on the production conditions and operation requirements. The traditional method relies on experiments and experience that are time-consuming and error-prone.

Until very recently, no efforts on the direction selection for the controller's action in the multiloop control system have been made. A new design method is investigated when we are building IDSCA [Intelligent Direction Selector for Controller's Action], an intelligent system to help designer choose the controller's action direction for the multiloop control systems [Rao, Jiang and Tsai, 1988]. By using the "action direction functions" and the "principle of equivalence", the approach can simplify the complexity of the prototype problem and help the designer to select the accurate direction of the controller's action. This method is relatively simple but effective for industrial process control projects. Besides these advantages, it indicates that the application of AI technique to the real world can stimulate and help the development of both AI and the related prototype problem.

Definition of Action Direction Function

Consider a simple (single loop) feedback control system in Figure 9.1. In this diagram, each of the components that constitute the control system is represented by a "block". For a simple process control system, there exist four basic blocks represented by transfer functions as follows:

1. controlled process, G_p
2. final control element (control valve), G_v
3. controller, G_c
4. measuring device, G_m

For the sake of convenience, we will assume that $G_m = 1$.

The knowledge base of IDSCA consists of the knowledge of the process, the valve and the controller. The knowledge of the process is about the characteristic of the process, which describes how the process output varies with the process input. The knowledge of the valve concentrates on the open-close type of the valve. The knowledge of the controller, for our purpose, is mainly described in terms of the direction of the controller's action.

In general, the action direction function for a block, namely R_b (b = c, v, p), is defined in terms of the input-output relation of this block. When the output of the block increases as its input becomes larger, the action direction function for this block is assigned as "positive", and vice versa.

For the controller, the action direction function, R_c, is defined as "positive" if the output signal P of the controller increases as the error signal E becomes larger, and vice versa.

The action direction function for the control valve, R_v, is determined based on production safety and operation conditions. Usually we decide whether to use an air-to-open (or electric-to-open) or air-to-close (or electric-to-close) valve to assure the security of the process and equipment when an accident occurs. The action direction function is "positive" for an air-to-open valve and "negative" for air-to-close valve.

The action direction function for the process block, R_p, is defined as "positive" if the process output Y increases when the manipulative variable (the output signal of the control valve) Q becomes larger, and vice versa. In the case of positive R_p, the characteristic of the process is referred to as "positive".

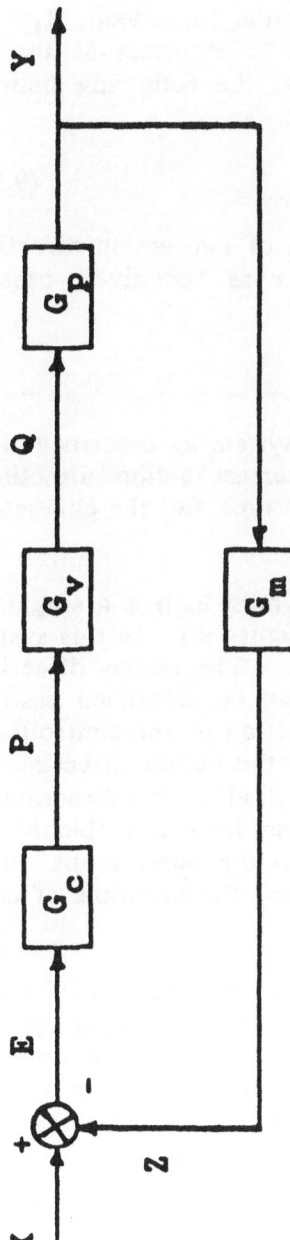

Figure 9.1 Block diagram of single loop control system

The action direction function for a loop, R_L, is defined as "positive" if the process output Y increases as the set point X increases. It can be shown that the following heuristic relation exists:

$$R_L = R_c * R_v * R_p \qquad\qquad (9.3.1)$$

where the sign product $R_1 * R_2$ of two action direction functions with identical signs is defined as "positive"; otherwise, it is "negative".

Principle of Equivalence

For a multiloop control system as described in Figure 9.2, our purpose is to select the correct action direction for all n controllers according to the valve type and the characteristics of all sub-processes in the system.

As we can see, the inner most loop is a single loop system, similar to that demonstrated in Figure 9.1. In this system, the only unknown block is the controller. The action direction functions for the process and the valve can be identified easily before the design procedure. Once the direction of the controller's action has been chosen, we may determine the action direction function for the loop R_L as the sign product of all action direction functions in the loop, and then treat this inner loop as a "block", which takes the place of the "valve block" in the outer loops. Based on the above methodology, we can develop the principle of equivalence as follows:

In the context of choosing the direction of the controller's action, a multiloop control system, consisting of n loops, is equivalent to a single loop system which is composed of controller 1 and process 1, both in the outer most loop (the first loop), together with a "valve block", the action direction function for which is the same as that for loop 2. That is,

$$R_{v1} = R_{L2} = R_{c2} * R_{v2} * R_{p2} \qquad\qquad (9.3.2)$$

In general, we have the following recursive equations:

$$R_{vi} = R_{L(i+1)} = R_{c(i+1)} * R_{v(i+1)} * R_{p(i+1)} \qquad\qquad (9.3.3)$$

in which i=1,2, ..., n-1.

111

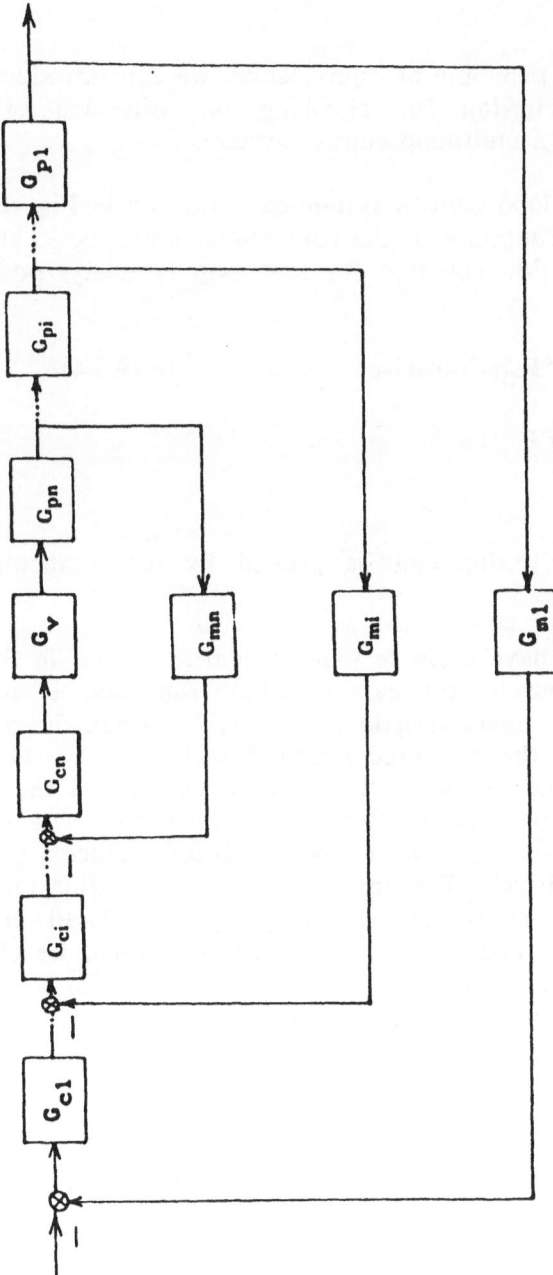

Figure 9.2 Block diagram of multiloop control system

Design Criterion

Based on the principle of equivalence, we can introduce the following design criterion for choosing the direction of the controllers' action in a multiloop control system.

For the multiloop control system as described in Figure 9.2, the action direction functions of the controllers should be so chosen that the action direction function for any loop is always positive. That is,

$$R_{Li} = R_{ci} * R_{vi} * R_{pi} = \text{"positive"} \qquad (9.3.4)$$

for i=1, 2, ..., n

Proof

The above criterion can be proved by the mathematical induction method.

For n=1, we have a single loop system as shown in Figure 9.1. Let us first consider the case in which the valve is air-to-open and the process characteristic is "positive". When the output variable Y increases, the measured signal Z will increase. For the regulator problem, the set point X is kept constant and the error signal E decreases. In order to compensate the increase in Y, the manipulative variable Q must be reduced since process characteristic is "positive". For an air-to-open valve, this requires the decrease of the controller output signal P. Based on the definitions mentioned above, a "positive action" should be chosen for the controller. The same conclusion can be drawn for the servo problem as well. Therefore, we can summarize this case as

Case 1: If R_v="positive" and R_p="positive",
then choose R_c to be "positive".

Three other possible cases can be similarly reasoned and the results are as follows:

Case 2: If R_v="positive" and R_p="negative",
then choose R_c to be "negative".

Case 3: If R_v="negative" and R_p="positive",
then choose R_c to be "negative".

Case 4: If R_v="negative" and R_p="negative",
then choose R_c to be "positive".

All of the above cases confirm the criterion, equation (9.3.4), for n=1.

Next, we need to show that the above criterion is valid for any (k + 1) loop system so long as it is satisfied for the k loop system. Here we recognize that k can be any integer between 1 and (n - 1).

For a (k + 1) loop system, we have the action direction function for the outer most loop (loop 1) as

$$R_{L1}=R_{c1}{}^*R_{v1}{}^*R_{p1} \qquad (9.3.5)$$

The principle of equivalence suggests that

$$R_{v1}=R_{L2}=R_{c2}{}^*R_{v2}{}^*R_{p2} \qquad (9.3.6)$$

Knowing that a sub-system with loops 2 to (k+1) constitutes a k loop system, we must have

$$R_{v1}=R_{L2}="positive" \qquad (9.3.7)$$

which is based on the premise that all k loop systems satisfy the criterion, equation (9.3.4).

This is equivalent to the single loop system with an air-to-open valve. Cases 1 and 2 above indicate that R_{L1} is always positive.

It can thus be induced that the criterion, equation (9.3.4), is valid for any multiloop system.

Recursion and Iteration

There are two ways to solve the problem. The first method is to start with the outer most loop and to end at the inner most loop. This leads to the use of recursion technique. The alternative method, here referred to as iteration technique, starts with the inner most loop and ends at the outer most loop. Recursion is a very powerful programming technique that often proves quite difficult to grasp initially and requires twice as many firings as the iterative version. Even though OPS5 is a Lisp-like programming tool that is very suitable for the use of recursion technique, recursion should be reserved for those problems that can not be easily solved in other ways.

In IDSCA, iteration technique is applied to solve the problem. It always starts with the inner most loop, in which the only unknown is the block of the controller. After receiving the information about the process characteristics and the valve type provided by the user, IDSCA determines the action direction of the controller in the inner most loop, and then treats the loop as a block as though it is the "valve" block in the outer next loop. In this way, the number of loops in the multiloop control system can be reduced by one for each cycle. Consequently, this method is repetitively applied until the controller's action direction in the outer most loop is determined.

Summary

A new approach to choosing the direction of controllers' action for the multiloop control systems is investigated when we are building an intelligent program for designing multiloop control systems. By using the "Action Direction Functions" and the "Principle of Equivalence", the approach can simplify the complexity of the prototype problem and help the designer to select the accurate direction of the controllers' action. Both recursion technique and iteration technique can be applied. The new design criterion is easy to be understood and applied in solving the complicated problems for the multiloop control systems.

The important significance behind the practical application of this new design criterion is that we realize that the development of intelligent systems for the real-world applications can stimulate and help the development of both AI and the related prototype problems. Finally, the reader should note that the methodology provided in this paper is based on the block diagram for the control system. For the instrumentation in which the definition of the error signal is reversed, the results for the direction of the controller's action are just opposite.

9.4 Adaptive feedback testing system

So far, many intelligent systems are developed to aid engineering designers in CAD processes. The design results of these systems are directly provided to the users as soon as the plausible solver is found. Generally speaking, the solver is only a candidate for a perfect design result.

The rule-based expert system often works in the domains where the conditions and conclusions are rarely certain. It picks the plausible solver as its answer to the prototype problem. In most cases, the imprecise knowledge may result in an error-prone solution even though the various skills of dealing with the imprecise knowledge are applied. In some expert systems, backtracking is utilized to test the reasoning process. However, the confidence in obtaining reliable solutions still cannot be guaranteed.

Once, Winston recounted [Bobcock, 1986] that MYCIN, an expert system for diagnosing medical cases [Shortlife, 1976], once prescribed a barrel of penicillin for a patient because of a data entry error. He noted "rule-based systems are real idiot savants" since they don't use models, don't use much experience and don't exhibit common sense. In the real-time environment, the reliability of design results directly affects manufacturing operation and product quality. How to enhance software reliability is the bottleneck for rule-based systems to be applied to the real engineering problems.

To solve the problems encountered above, an intelligent methodology is proposed to avoid using an error-prone solver as a design result. During the process of developing IDSCA [Rao, Jiang and Tsai, 1988], we develop an automatic test system to evaluate the candidate solver such that a correct final decision can be made to assure the reliability of the conclusions. This is very similar to the human intelligent behavior that a human expert tests his result to assure the correctness of the calculation when he faces a plausible solution. Such a technique is based on intelligence, not on knowledge. Nowadays, it is very needed for the intelligent CAD commercially.

IDSCA possesses such an intelligent feature that it can test its design results automatically and modify the design program adaptively if the obtained result is not reasonable. The configuration, namely Adaptive Feedback Testing System (AFTS), is described in Figure 9.3. AFTS consists of four models, each of which performs a specific function. The managing model controls the selection, operation and communication in AFTS, and the distributing model delivers the design results of IDSCA to the testing model or modifying model whenever an improvement is needed.

Through AFTS, each of the IDSCA's design results is fed back to the prototype plant. First of all, the safety factor is tested against the various possible accidents. If the design results can guarantee the security of production (i.e., no conflict exists in the design results), then such factors as operation requirements and production conditions are tested in turn. The function of AFTS is based on the following heuristic laws:

Law I: If no conflict exists, send the obtained design results to the user as the final decision.

Law II: If any conflict is found, modify the design results currently being designed, and then feed the modified results back to the plant for further testing.

Law III: If Law II always fails, report the conflict factors to user.

With the above three laws governing the execution of AFTS, even though each individual loop designed is in the different case that may be very sophisticated and likely to change for one reason or another, AFTS can always take the suitable routine to handle the particular situation. The strategy of self-testing and self-modifying makes IDSCA possess an adaptive capability, or in other words, a learning behavior to deal with the complicated problems.

The production rules in AFTS are organized into several rule bases, which includes testing rule base, redesign rule base and interface rule base. The inference cycle of AFTS is controlled by a set of meta-rules, which determine when a rule base has completed its work and what rule base is to invoke next. The structure for meta-level reasoning and meta-level control and the application of AFTS are presented in Chapter 2. With its universal mechanism, Adaptive Feedback Testing can be applied to other knowledge-based CAD systems.

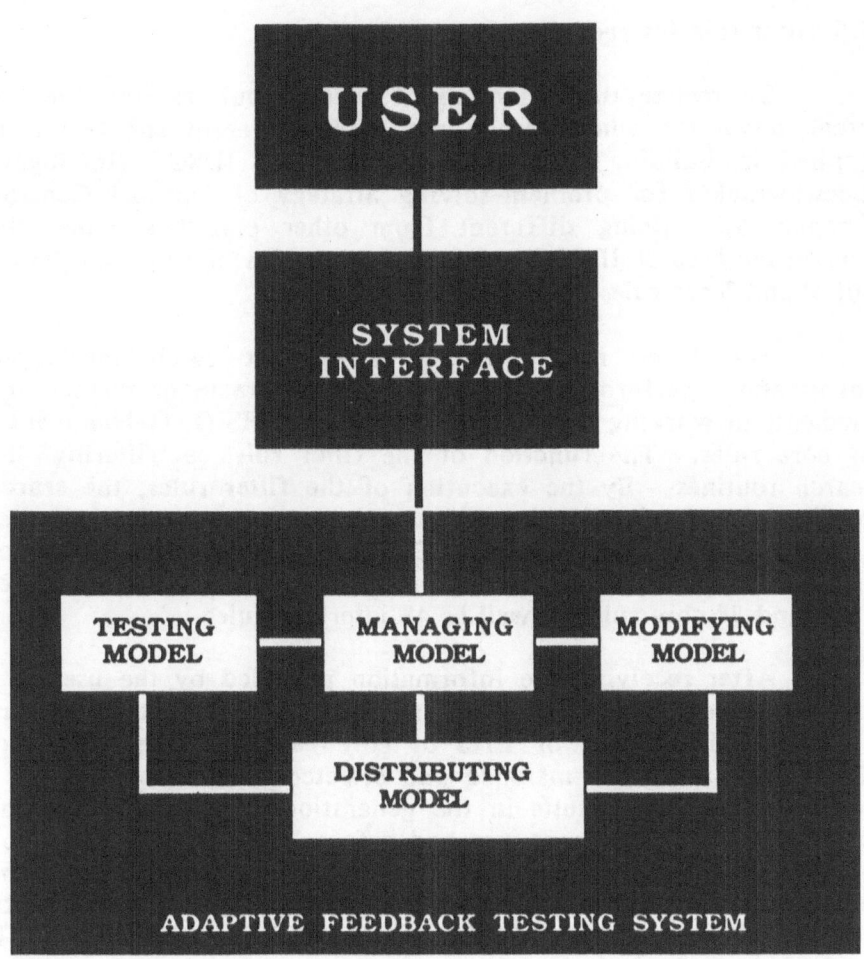

Figure 9.3 Adaptive Feedback Testing System

9.5 Filter rule for reducing the search of rules

To realize the reduction of search routines, the idea to break down the whole target into many different sub-targets is applied in building the knowledge base for IDSOC (Intelligent Decisionmaker for problem-solving Strategy of Optimal Control, Chapter 5). Being different from other expert systems, the knowledge base of IDSOC consists of two types of rules: the "filter rules" and "core rules".

The filter rules are the rules that catch the input information, perform heuristic search, and create or change the elements in working memory which are the LHS (Left-Hand-Side) of core rules. The function of the filter rules is "filtering" the search routines. By the execution of the filter rules, the search routines can be simplified greatly and the needed production rules can be reduced by about 30%. Currently, the knowledge base of IDSOC consists of 98 production rules, in which there are 16 filter rules and 34 core rules as well as 48 interface rules.

After receiving the information provided by the user, the input interface rules initiate data such that LHS of filter rules are formed, and the part of LHS of core rules may be generated. Then, filter rules are matched and selected. The execution of a selected filter rule results in the generation of the LHS of some core rules which were not matched before. Through the execution of the core rules, the structure of a solution is constructed. The output interface rules send the solver to the user. The inference engine will not stop until the complete output information is generated. IDSOC was implemented by an expert system building tool, OPS5 [Brownston, et al., 1985], on a VAX computer. Its reasoning process is demonstrated by Figure 9.4.

The concept of the filter rule can be applied to other knowledge base systems. Its structure advantage allows us to organize the knowledge bases better for intelligent systems. It provides a means to improve the efficiency of knowledge-based systems.

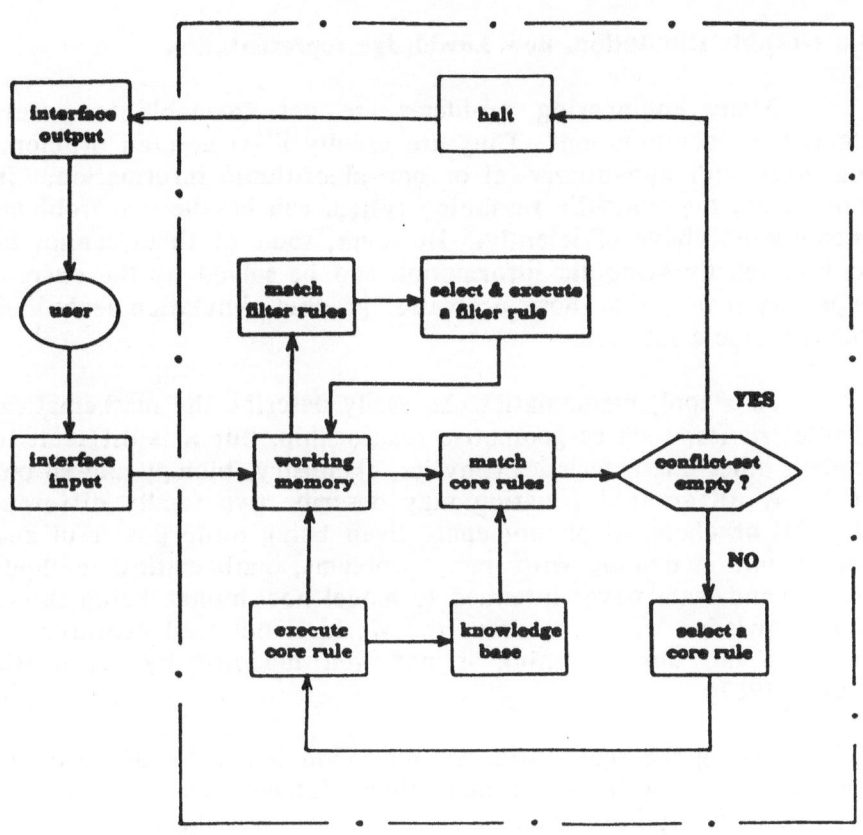

Figure 9.4   Reasoning process of IDSOC

9.6 Graphic simulation: new knwoledge representation

Many engineering problems are not amenable to purely algorithmic computation. They are usually ill-structured problems that deal with non-numerical or non-algorithmic information. In most cases, the symbolic reasoning system can handle the problems encountered above efficiently. However, some of them cannot be represented by symbolic information and be solved by the current expert systems. For those problems, graphic simulation technique may provide a solution.

As a tool, mathematics can easily describe the mathematical characteristics, such as geometric relationship, but it is difficult to explain other characteristics (physics, chemistry, biology and so on) well. A differential equation may describe two totally different physical or chemical phenomena. Even being quite powerful and appropriate in dealing with many problems, mathematical methods are not and were never intended to model how human being think. The knowledge we know about the world is not well captured by numbers and our reasoning is not well modeled by arithmetic [Davis, 1987].

Facing the real world, we may find that a lot of problems are solved very well even though their mathematical expression is not available. It is by using rules of thumb, not the mathematical understanding, that the engineers can solve the non-algorithmic problems and process the non-numeric information. Many artificial intelligence scientists are working to solve these problem. Expert systems, computer vision and natural language understanding are the main fields of AI research, which deal with simulating and modelling how the human being think and reason. Advance in these fields provides the hope that in the future, visual perception will be performed by the computers.

As a useful technique, graphics has been extensively used in simulation to help facilitating model construction and debugging, providing interactive control during the running of the simulation, and animating the simulation to help the user visualize the results [Shannon, 1986].

Graphic simulation can be utilized as a new means to represent knowledge and to describe the real world engineering problems more clearly and perfectly. Usually, graphic simulation is applied to qualitative analysis, but it can also be applied to quantitative simulation. The following two case studies will illustrate that besides numeric and symbolic information, graphic information can serve as a fundamental and powerful way to represent the knowledge about the real world.

Case study 1: solution of differential-integral equation

A differential-integral equation is given as:

$$c^*X = n \int [1 + Y'^2]^{(n/2-1)} dY' \qquad (9.6.1)$$

where, $Y' = dY/dX$.

This equation is very difficult to solve. Its solutions can be represented by several different types of geometric curves, depending on the value of parameter n. Parameter c only affects the shapes of the curves. For example, for n=-1, we get the elliptic curve. For n=-2, the curve is a pendulum. A parabolic curve corresponds to the case in which n=2. When solving this equation, symbolic integration is hardly applicable because of the complexity of the problem-solving procedure, especially for the cases with a large n. On the other hand, numeric information cannot represent the solution clearly even though the problem can be solved by numeric computation. Graphic simulation can handle the problem encountered above easily and animate the computation results such that we can easily recognize geometric curves.

Case study 2: vision description

Consider a scenario that you are recommending a lady to work as a TV model. If you only provide numeric information about this lady, such as height, weight, age, and so on, it is still difficult to build a basic impression of the lady's looks. However, a simple but effective description can be easily made with symbolic information. Nevertheless, the best impression can only be made by a picture or vision representation.

In many applications, graphic simulation can provide not only qualitative but also quantitative descriptions for knowledge representation. For those problems that numeric computation and symbolic reasoning are not appropriate to be used, graphic simulation can offer a useful alternative. Such type of vision representation is an attempt to overcome shortcomings of symbolic reasoning and numeric computation.

In summary, graphic simulation represents both qualitative and quantitative information, which is essential to solving many engineering problems. It provides us a new means to describe the real world and helps to develop more intelligent expert systems. Besides computer vision and image processing, graphic simulation provides us another means to describe the real world with easily conceived graphics. This will stimulate us to build "more intelligent" systems.

6.7 Integration system for distributed AI

The development of the integrated intelligent system was motivated by the fundamentals of both large scale system and distributed control. It contributes to the theory and technology development for the distributed artificial intelligence and software engineering. Also, several research topics are generated from the integrated intelligent system, in which conflicting reasoning, parallel computing for intelligent system, and sampling expertise will pave the way for us to build theoretic aspects and applications for intelligent system as well as the related engineering domains.

An integrated intelligent system is a large knowledge integration environment which consists of several symbolic reasoning systems and numerical computation packages. These computer programs may be written in different AI languages or tools or various conventional languages, and be used independently. They are under the control of a supervising expert system, namely, meta-system. Detailed description about integrated intelligent system has been presented in the previous chapters, where the interested readers can find more information. Here, I only briefly describe system integration architecture, and list the functions of meta-system.

INTEGRATION ARCHITECTURE

1. integration of the knowledge of different disciplinary domains;
2. integration of empirical expertise and analytical knowledge;
3. integration of various different objectives, such as research and development, engineering design and implementation, process operation and control;
4. integration of different expert systems;
5. integration of different numerical packages;

6. integration of symbolic processing systems and numerical computation packages;

7. integration of different information, such as symbolic, numeric and graphic information.

MAIN FUNCTIONS OF META-SYSTEM

1. Meta-system is the coordinator to manage all symbolic reasoning systems and numerical computing routines in an integrated intelligent system.

2. Meta-system distributes knowledge source into separate expert systems and numeric routines so that integrated intelligent system can be managed effectively, and knowledge bases of expert systems are easier to be modified by commercial users other than their original developers.

3. Meta-system is an integrator which can help us acquire new knowledge easily.

4. Meta-system can provide an optimal solution for conflicting results and events among different expert systems.

5. Meta-system may provide the possibility of parallel processing in integrated intelligent system.

6. Meta-system can communicate with measuring devices and final control elements in the control systems and transform various nonstandard input/output signals into standard communication signals.

Meta-system has been implemented with three different software environments (OPS5 and C as well as TURBO-PROLOG). It has been applied to many disciplinary fields, such as intelligent process control, computer integrated manufacturing systems, maintenance support system for air-traffic control, and conventional software development. As a universal architecture, the integrated intelligent system has attracted significant attention from both academia and industry, and is expected to lead to a new era for the applications of artificial intelligence techniques to the real-world engineering and scientific problems.

10 Conclusions

In summary, an integrated intelligent system for real time control technology has been successfully developed, which can be described in the following:

1. We have successfully applied AI techniques to real time control problems. Intelligent control is an interdisciplinary field, which extensively applies the knowledge of computer science and artificial intelligence as well as system engineering to industrial processes.

2. Knowledge-based systems can help us solve ill-structured problems in the real time control. The knowledge-based systems are the additional welcome to the existing knowledge. They can never be the replacement for today's science and technology in real time control.

3. The Integrated Intelligent System is an innovative intelligent environment, which can help us achieve knowledge integration and management. The Integrated Intelligent System is a universal configuration that will lead a new era for AI applications to the real-world problems.

4. New knowledge may be generated in the processes of developing knowledge-based systems, which may complement the knowledge of artificial intelligence techniques, control engineering, and related domains.

Finally, as we know that knowledge acquisition usually is very difficult. Even with the help of knowledge engineers, some private knowledge may still be difficult to be transferred. Training some domain experts into intelligence engineers is important and necessary task to extend AI applications to the real-world.   Intelligence engineers are different with knowledge engineers. Originally, they are the experts working in the related application domains.   Through a comparatively short period of training, they learn the fundamentals of applied artificial intelligence techniques, get the hand-on experience on developing knowledge-based systems. Then, they could build up much better knowledge-based systems to solve their domain problems, and extend AI applications successfully.

References

Akrif, O. and G.L. Blankenship, 1987. "Computer Algebra for Analysis and Design of Nonlinear Control Systems," Proc. of American Control Conference, Atlanta, GA, pp. 547-554.

Asbjornsen, O.A., 1984. "Challenges in Modern Process Control", Computers & Chemical Engineering. 8, pp. 275-284.

Astrom, K.J., 1985. "Process Control -- Past, Present, and Future," IEEE Control Magazine. August, pp. 3-10.

Astrom, K.J., J.J. Anton and K.E. Arzen, 1986. "Expert Control," Automatica. 22, pp. 277-286.

Astrom, K.J., P. Hagander, P. and J. Sternby, 1984. "Zeros of Sampled Systems," Automatica. 20, pp. 31-38.

Astrom, K.J., and B. Wittenmark, 1974. "On Self-Tuning Regulators," Aotmatica. 19, pp. 185-199.

Babcock, C., 1986. "MIT Director: AI Most Expand Problem Models," Computerworld, November 24, pp. 20.

Bourne, D.A., 1986. "CNL A Meta-Interpreter for Manufacturing," AI Magazine. 7(4), pp. 86-96.

Bowerman, R.G. and D.E. Glover, 1988. Putting Expert Systems into Practices, Van Nostran Reinhold Company, New York.

Brownston, L., R. Farrell, E. Kant and N. Martin, 1985. Programming Expert Systems in OPS5, Reading MA: Addison-Wesley.

Buchanan, B.G., 1985. "Expert Systems" in "An Overview of Automated Reasoning and Related Fields," J. Automat. Reasoning. 1, pp. 28-34.

Burg, B., L. Foulloy, J.C. Heudin and B. Zavidovique, 1985. "Behaviour Rule Systems for distributed Process Control," Proc. IEEE 2nd Conf. AI Applic., Miami, FL, pp. 198-203.

Butler, C.W., E.D. Hodil and G.L. Richardson, 1988. "Building Knowledge-Based Systems with Procedural Languages," IEEE Expert, Summer, pp. 47-59.

Buxton, B., 1985. "Impact of Packaged Software for Process Control on Chemical Engineering Education & Research," Chem. Eng. Education. 19(2), pp. 144-147.

Cha, J.Z., M. Rao, J. Zhou and Z.C. Shi, 1990. "Integrated Intelligent Software Systems and Their Applications in Mechanical Engineering", Journal of Tianjin University, 90(2), pp. 44-51.

Cha, J., Zhao and M. Rao, 1991. "Integrated Intelligent System for Gear Manufacturilng," (submitted for publication) Applied Artificial Intelligence.

Clark, D.W., 1984. "Self-Tuning Control of Nonminimum-Phase Systems," Automatica. 20, pp. 501-517.

Clancey, W.J., 1987. Knowledge-Based Tutoring, The MIT Press, Cambridge, MA.

Colgren, R.D., 1988. "Development of a Workstation for the Integrated Design and Simulation of Flight Control Systems," Proc. NAECON 88, Dayton, OH, pp. 380-387.

Cruise, A., R. Ennis, A. Finkel, J. Hellerstein, D. Klein, D. Loeb, M. Masullo, K. Milliken, H. Van Woerkom, N. Waite, 1987. "YES/L1: Integrating Rule-Based, Procedural, and Real-Time Programming for Industrial Applications," Proc. IEEE Third AI Application Conf., pp. 134-139.

Davis R., 1987. "Knowledge-Based Systems: The View in 1986," in AI in the 1980s and Beyond, (Grimson and Patil eds.), MIT Press, pp. 16-19.

Davis R. and D. Lenat, 1982. Knowledge-Based Systems in Artificial Intelligence, New York, NY: McGraw-Hill, Inc.

Edgar, T.F., 1987. "Current Problems in Process Control", IEEE Control Systems Magazine, April, pp. 13-15.

Eldeib, H.K. and S. Tsai, 1988. "Applications of Symbolic Manipulation in Control System Analysis and Design," Third IEEE Intern. Symposium on Intelligent Control, Arlington, Virginia.

127

Fu, K.S., 1971. "Learning Control Systems and Intelligent Control Systems: An Intersection of Artificial Intelligence and Automatic Control", IEEE Trans. Automat. Contr., AC-16, pp. 70-72.

Gagnepain, J.P. and D.E. Seborg, 1982. "Analysis of Process Interactions with Applications to Multiloop Control System Design," Ind. Eng. Chem. Process Des. Dev., 21, pp. 5-11.

Gallanti, M., G. Guida, L. Spampinato and A. Stefanini, 1985. "Representing Procedural Knowledge in Expert Systems: An Application to Process Control," Proc. Intern. Joint Conf. on AI, 85, pp. 346-3352.

George, M.P. and O. Firschein, 1985. "Expert Systems for Space Station Automation", IEEE Control System Magazine, November, pp. 3-8.

Goodwin, G.C., and K.S. Sin, 1981. "Adaptive Control of Nonminimum-Phase Systems," IEEE Trans. Autom. Control, 26, pp. 478-482.

Handelman, D.A. and R.F. Stengel, 1987. "An Architecture for Real-Time Rule-Based Control," Proc. American Control Conference, Minneapolis, MN, pp. 1636-1642.

Hayes-Roth, B., 1985. "A Blackboard Architecture for Control," Artificial Intelligence 26, pp. 251-321.

Ionescu, D. and I. Trif, 1988. "Hierarchical Expert System for Process Control," Proc. American Control Conference, Atlanta, GA, pp. 1-6.

Inselberg, A., 1985. "Intelligent Instrumentation & Process Control," Proc. IEEE 2nd Conf. AI Applic., Miami, FL, pp. 302-306.

Ishii, K. and S. Hayami, 1988. "Expert Systems in Japan," IEEE Expert, Summer, pp. 69-74.

Jacobstein, N., C.T. Kitzmiller and J.S. kowalik, 1988. "Integrating Symbolic and Numeric Methods in Knowledge-Based Systems: Current Status, Future Prospects, Driving Events," Coupling Symbolic and Numerical Computing In Expert Systems, II (ed. by J. S. Kowalik and C. T. Kitzmiller), pp. 3-11, New York, NY: Elsevier Science Publishers B.V.

James, J.R., 1987. "A Survey of Knowledge-Based Systems for Computer-Aided Control System Design," Proc. of American Control Conference, Minneapolis, MN, pp. 2156-2161.

James, J.R., D.K. Frederick, P.P. Bonissone and J.H. Taylor, 1985. "A Retrospective View of CACE-III: Considerations in Coordinating Symbolic and Numeric Computation in a Rule-Based Expert System," Proc. IEEE 2nd Conf. AI Applic., Miami, FL, pp. 532-538.

Johnston, R.D. and G.W. Barton, 1984. "Improved Process Conditioning using Internal Control Loops," Int. J. Control. 40, pp. 1051-1063.

Jones, M., 1985. "Applications of Artificial Intelligence within Education," Comp. & Math. with Appls., 11(5), pp. 517-526.

Jury, E,I., 1961. IRE Trans. Autom. Control, 6, p. 324; 1964, Theory and Application of Z Transforms, (New York: Wiley). pp. 84-141.

Kitzmiller, C.T. and J.S. Kowalik, 1987. "Coupling Symbolic and Numeric Computing in Knowledge-Based Systems," AI Magazine. Summer, pp. 85-90.

Korn, G.A., and Korn, T.M., 1961. Mathematical Handbook for Scientists and Engineers New York: Mcgraw-Hill, p. 16.

Lamont, G. and M.W. Schiller, 1987. "The Role of Artificial Intelligence in Computer-Aided Design of Control Systems," Proc. IEEE 26th Conf. on Decision and Control, Los Angles, CA, pp. 1960-1965.

Lewin, D.R. and M. Morari, 1987. "ROBEX: An Expert System for Robust Control Design," Proc. American Control Conf., Minneapolis, MN.

Lirov, Y., E.Y. Rodin, B.G. McElhaney and L.W. Wilbur, 1988. "Artificial Intelligence Modelling of Control Systems," Simulation 50, pp. 12-24.

Lizza, C. and C. Friedlander, 1988. "The Pilot's Associate: A Forum for the Integration of Knowledge Based Systems and

Avionics," Proc. National Aerospace and Electronics Conf. 88, Dayton, OH, pp. 1252-1258.

Lu, S.C.-Y., 1989. "Knowledge Processing for Engineering Automation," Proc. 15th Conf. on Production Research and Technology, Berkeley, CA, pp. 455-468.

Luxhoj, J.T. and M.S. Jones, 1988. "A Computerized population Model for System Repair/Replacement," Computer and Industrial Engineering, 14(3), pp. 345-359.

MacFarlane, A.G.J., 1972. "A Survey of Some Resent Results in Linear Multivariable Feedback Theory," Automatica, 8, pp. 455-492.

Moore, R.L., L.B. Hawkinson, C.G. Knickerbrocker and L.M. Churchman, 1984. "A Real-Time Expert System for Process Control," Proc. IEEE 1st AI Applic. Conf., Denver, CO, pp. 569-576.

Moray, N., 1986, "Operator Models in Process Control", Proc. IEEE Systems, Man, and Cybernetics, pp. 247-251.

Moses, J., 1967. "Symbolic Integration," PhD Thesis, Massachusetts Institute of Technology, Cambridge, MA.

Okuno, H.G. and A. Gupta, 1988. "Parallel Execution of OPS5 in QLISP," Proc. IEEE AI Application Conf., pp. 268-273.

Orelup, M.F. and P.R. Cohen, 1988. "Dominic II: Meta-Level Control in Iterative Redesign," Proc. American Control Conf., Atlanta, GA, pp. 25-30.

Oren, T.I. and B.P. Zeigler, 1988. "Artificial Intelligence in Modelling and Simulation: Directions to Explore," Simulation 50, pp. 131-134.

Owens, D.H., 1981. "Some Unifying Concepts in Multivariable Feedback Design," Int. J. Control, 33, pp. 701-711.

Pang, G.K.H, M. Vidyasagar and A.J. Heunis, 1990. "Development of a New Generation of Interactive CACSD Environments," IEEE Control Systems Magazine, 10(5), pp. 40-44.

Pang, G.K.H and A.G.J. MacFarlane, 1987. An Expert System Approach to Computer-Aided Design of Multivariable Systems, Springer-Verlag, New York.

Parunak, H.V.D., B.W. Irish, J. Kindrick and P.W. Lozo, 1985. "Fractal Actors for Distributed Manufacturing Control", Proc. IEEE 2nd AI Applications Conf., Miami, FL, pp. 653-660.

Qian, D. and Y. Lu, 1987. "Applications of Fuzzy Set Theory to Knowledge Representation and Reasoning in Dynamic Systems," Proc. IEEE Intern. Conf. on System, Man, and Cybernetics, Alexandria, VA, pp. 1032-1035.

Rao, M., J.Z. Cha, and J. Zhou, 1990. "New Software Platform for Intelligent Manufacturing", Proc. AAAI 90 Workshop on Intelligent Manufacturing Architecture, Boston, MA, pp. 62-65.

Rao, M., J.Z. Cha, J. Zhou, 1991. "PHIIS: Parallel Hierarchical Integrated Intelligent System for Engineering Design Automation", Engineering Applications of Artificial Intelligence, 4.

Rao, M., C. Theisen and J. Luxhoj, 1990. "Intelligent System for Air-Traffic Control", 5th IEEE Intern. Symp. on Intelligent Control, Philadelphia, PA, pp. 859-863.

Rao, M. and T.S. Jiang, 1989. "Expert Systems for Process Control: A Survey," Proc. Instrument Society of American Southeastern Conf. '89, Pensacola, FL.

Rao, M. and T.S. Jiang, 1989. "Expert-Aided Process Control Laboratory," Intern. J. Applied Eng. Education, 6(2), pp. 227-231.

Rao, M., T.S. Jiang and J.P. Tsai, 1990. "Integrated Intelligent Simulation Environment," Simulation, 54(6), pp. 291-295.

Rao, M., T.S. Jiang, and J.P. Tsai, 1989. "Combining Symbolic and Numerical Processing for Real-Time Intelligent Control", Eng. Applic. of Artificial Intelligence, 2, pp. 19-27.

Rao, M. and T.S. Jiang, 1988. "Simple Criterion for Testing Non-Minimum-Phase Systems," Intern. J. Control, 47, pp. 653-656.

Rao, M., J.P. Tsai and T.S. Jiang, 1988. "An Intelligent Dicisionmaker for Optimal Control," Applied Artificial Intelligence, 2, pp. 285-305.

Rao, M., T.S. Jiang and J.P. Tsai, 1988. "IDSCA: An Intelligent Direction Selector for the Controller's Action in Multiloop Control Systems," Intern. J. Intelligent Systems, 3, pp. 361-379.

Rao, M., T.S. Jiang and J.P. Tsai, 1988. "Integrated Architecture for Intelligent Control," Third IEEE International Symposium on Intelligent Control, Arlington, Virginia, pp. 81-85.

Rao, M., J.P. Tsai and T.S. Jiang, 1987. "Adaptive Feedback Testing System: Case Studies and Application," Mid-West Symp. Artificial Intelligence & Cognitive Science, Chicago, IL, pp. 43-50.

Rao, M., J.P. Tsai and T.S. Jiang, 1987. "A Framework of Integrated Intelligent Systems," Proc. IEEE Intern. Conf. on System, Man, and Cybernetics, Alexandria, Virginia, pp. 1133-1137.

Rao, M., X. Zheng and T.S. Jiang, 1987. "Graphic Simulation: Beyond Numerical Computation and Symbolic Reasoning," Proc. IEEE Intern. Conf. Systems, Man, and Cybernetics, Beijing, China, pp. 523-524.

Rao, M. Y. Ying and J. Corbin, 1991. "Intelligence Engineering Approach to Pulp and Paper Process Control", 91th CPPA Annual Conf., Montreal, Quebec, pp. 195-200.

Rembold, U., C. Blume and R. Dillmann, 1985. Computer Integrated Manufacturing Technology and Systems. Marcel Deker, Inc. New York.

Roat, S.D. and S.S. Melsheimer, 1987. "Microcomputer-Aided Control Systems Design," Chemical Engineering Education, Winter, pp. 34-39.

Roboam. M., K. Sycara and M. Fox, 1990. "The Intelligent Networking Architecture: A Tool for Manufacturing Enterprise Integration," Proc. AAAI'90 Workshop on Intelligent Manufacturing Architectures, pp. 1-4.

Rosenbrock, H.H., 1969. "Design of Multivariable Control Systems using the Inverse Nyquist Array," Proc. Inst. Elect. Engrs. 116, pp. 1929.

Sage, A.P. and C.C. White, III, 1977. Optimum Systems Control, 2nd ed., Englewood Cliffs, NJ: Prentice-Hall.

Sang, Z.T., M. Rao and T.W. Weber, 1986. "A Microcomputer-Based Simulation Laboratory for Process Control," Proc. SCS Multiconference. Application of Microcomputer Simulation, San Diego, CA, pp. 213-218.

Saridis, G.N., 1983. "Intelligent Robotic Control", IEEE Trans. Automatic Control. AC-28, pp. 547-557.

Saridis, G.N. and K.P. Valavanis, 1988. "Analytical Design of Intelligent Machines," Automatica. 24, pp. 123-133.

Shannon, R., R. Mayer and H. Adelsberger, 1985. "Expert Systems and Simulation," Simulation 44, no. 6, pp. 275-284.

Shannon, R. 1986. "Intelligent Simulation Environment," Proc. Intelligent Simulation Environment, San Diego, CA, pp. 150-156.

Shirley, R.S., 1987. "Some Lessons Learned Using Expert Systems for Process Control," Proc. American Control Conf., pp. 1342-1346.

Shortlife, E.H., 1976. MYCIN: Computer-based Medical Consultations, New York, NY: Elsevier.

Slagle, J.R., 1963. "A Heuristic Program That Solves Symbolic Integration Problems in Freshman Calculus," JACM. 10, pp. 507-520.

Stephanopoulos, G., 1984. Chemical Process Control. Prentice-Hall, Englewood Cliffs, NJ.

Stephanopoulos, G., 1986. "Expert Systems in Process Control," Proc. 3rd Chem. Process Control Conf., pp. 803-806.

Stephanopoulos, G., 1987. "The Future of Expert Systems in Chemical Engineering," Chem. Eng. Progress. Sept., pp. 44-51.

Su, S.Y.W. and Lam, 1990. Object-Oriented Knowledge Base Management Technology for improving productivity and Competitiveness in Manufacturing," NSF Grantees Conf. on Design and Manufacturing Systems, Tempe, AZ.

Talukdar, S.N., E. Cardozo and L.V. Leao, 1986. "Toast: Power System Operator's Assistant," IEEE Computer July, pp. 53-60.

Taylor, J.H. and D.K. Frederick, 1985. "An Expert System Architecture for Computer-Aided Control Engineering," IEEE Proceedings, 72, pp. 1795-1805.

Thoma, M., 1962. "Ein einfaches Verfahren zur Stabilitatsprufung von linear Abtastsystemen," Regelungstechnik, 10, p. 302; 1963, Ibid., 11, p. 70.

Thompson, T. and R. Wojcik, 1984. "MELD: An Implementation of a Meta-Level Architecture for Process Diagnosis," Proc. IEEE First AI Applications Conf., pp. 321-330.

Tzouanas, V., C. Georgakis, W. Luyben and L. Ungar, 1988. "Expert Multivariable Control," Comput. & Chemical Engineering.

Wang, J.C., 1980. Chemical Process Control Engineering, Chemical Industrial Press, Beijing.

Watanabe, K., Y. Ishiyama and M. Ito, 1983. "Modified Smith Predictor Control for Multivariable Systems with Delays and Unmeasurable Step Disturbances," Int. J. Control, 37, pp. 959-973.

Winston, P.H. 1984. Artificial Intelligence, 2nd ed. Reading, MA: Addison-Wesley.

Wright, M.L., M.W. Green, G. Fiegl and P.F. Cross, 1986. "An Expert System for Real-Time Control," IEEE Software, March, pp. 16-24.

Wong, F.S., W. Dong and M. Blanks, 1988. "Coupling of Symbolic and Numerical Computations on a Microcomputer," AI in Engineering, 3, pp. 32-38.

Xi, Y., and Schmidt, G., 1985. "A Note on the Location of the Roots of a Polynomial," IEEE Trans. Autom. Contr., 30, pp. 78-80.

Yeomans, R.W., A. Chaudry and P.J. Tenhagen, 1985. Design Rules for A CIM System, North-Holland.

Zadeh, L.A., 1978. "Fuzzy Sets as a Basis for a Theory of Possibility," Fuzzy Sets and Systems, 1, pp. 3-28, 1978.

Zeigler, B.P., 1984. "Multifaceted Modeling Methodology: Grappling with the Irreducible Complexity of Systems," Behavioral Science 29, pp. 169-178.

Lecture Notes in Control and Information Sciences

Edited by M. Thoma and A. Wyner

Lecture Notes in Control and Information Sciences

Edited by M. Thoma and A. Wyner

Lecture Notes in Control and Information Sciences

Edited by M. Thoma and A. Wyner